MATA REFERENCE MANUAL
VOLUME 1
[M-0] – [M-3]
RELEASE 10

A Stata Press Publication
StataCorp LP
College Station, Texas

Stata Press, 4905 Lakeway Drive, College Station, Texas 77845

ISBN-10: 1-59718-037-8 (volumes 1–2)
ISBN-10: 1-59718-035-1 (volume 1)
ISBN-10: 1-59718-036-X (volume 2)
ISBN-13: 978-1-59718-037-5 (volumes 1–2)
ISBN-13: 978-1-59718-035-1 (volume 1)
ISBN-13: 978-1-59718-036-8 (volume 2)

The suggested citation for this software is

StataCorp. 2007. *Stata Statistical Software: Release 10*. College Station, TX: StataCorp LP.

Brief Contents

Volume 1

Volume 2

Table of Contents

Contents of Mata Volume 1

[M-3] Commands for controlling Mata

Contents of Mata Volume 2

[M-4] Index and guide to functions

[M-5] Mata functions

[M-6] Mata glossary of common terms

Cross-Referencing the Documentation

When reading this manual, you will find references to other Stata manuals. For example,

[U] **26 Overview of Stata estimation commands**

[R] **regress**

[D] **reshape**

The first is a reference to chapter 26, *Overview of Stata estimation commands* in the *Stata User's Guide*, the second is a reference to the `regress` entry in the *Base Reference Manual*, and the third is a reference to the `reshape` entry in the *Data Management Reference Manual*.

All the manuals in the Stata Documentation have a shorthand notation:

[GSM]	*Getting Started with Stata for Macintosh*
[GSU]	*Getting Started with Stata for Unix*
[GSW]	*Getting Started with Stata for Windows*
[U]	*Stata User's Guide*
[R]	*Stata Base Reference Manual*
[D]	*Stata Data Management Reference Manual*
[G]	*Stata Graphics Reference Manual*
[P]	*Stata Programming Reference Manual*
[XT]	*Stata Longitudinal/Panel-Data Reference Manual*
[MV]	*Stata Multivariate Statistics Reference Manual*
[SVY]	*Stata Survey Data Reference Manual*
[ST]	*Stata Survival Analysis and Epidemiological Tables Reference Manual*
[TS]	*Stata Time-Series Reference Manual*
[I]	*Stata Quick Reference and Index*
[M]	*Mata Reference Manual*

Detailed information about each of these manuals may be found online at

http://www.stata-press.com/manuals/

[M-0] Introduction to the Mata manual

Title

[M-0] intro — Introduction to the Mata manual

Contents

Section	Description
[M-1]	**Introduction and advice**
[M-2]	**Language definition**
[M-3]	**Commands for controlling Mata**
[M-4]	**Index and guide to functions**
[M-5]	**Functions**
[M-6]	**Mata glossary of common terms**

Description

Mata is a matrix programming language that can be used by those who want to perform matrix calculations interactively and by those who want to add new features to Stata.

This entry describes this manual and what has changed since Stata 9.

Remarks

This manual is divided into five sections. Each section is organized alphabetically, but there is an introduction in front that will help you get around.

If you are new to Mata, here is a helpful reading list: start by reading

[M-1] **first**	Introduction and first session
[M-1] **interactive**	Using Mata interactively
[M-1] **how**	How Mata works

You may find other things in section [M-1] that interest you. For a table of contents, see

[M-1] **intro**	Introduction and advice

Whenever you see a term that you are unfamiliar with, see

[M-6] **glossary**	Mata glossary of common terms

3

Now that you know the basics, if you are interested, you can look deeper into Mata's programming features:

[M-2] **syntax**	Mata language grammar and syntax

[M-2] **syntax** is pretty dense reading, but it summarizes nearly everything. The other entries in [M-2] repeat what is said there but with more explanation; see

[M-2] **intro**	Language definition

Along the way, you will eventually be guided to sections [M-4] and [M-5]. [M-5] documents Mata's functions; the alphabetical order makes it easy to find a function if you know its name but makes learning what functions there are hopeless. That is the purpose of [M-4]—to present the functions in logical order. See

[M-4] **intro**	Index and guide to functions
Mathematical	
[M-4] **matrix**	Matrix functions
[M-4] **solvers**	Matrix solvers and inverters
[M-4] **scalar**	Scalar functions
[M-4] **statistical**	Statistical functions
[M-4] **mathematical**	Other important functions
Utility and manipulation	
[M-4] **standard**	Functions to create standard matrices
[M-4] **utility**	Matrix utility functions
[M-4] **manipulation**	Matrix manipulation functions
Stata interface	
[M-4] **stata**	Stata interface functions
String, I/O, and programming	
[M-4] **string**	String manipulation functions
[M-4] **io**	I/O functions
[M-4] **programming**	Programming functions

What's new

Mata was first released in Stata 9, and thus the release of Stata 10 amounts to Mata's first update. Here is what's changed and what's new:

1. New Stata command `include` is a variation on `do` and `run` and is useful for implementing `#include` for header files in advanced programming situations. See [P] **include** and type `viewsource optimize.mata` for an example of use.

2. Mata now has structures, which will be of special interest to those writing large systems. See [M-2] **struct** and [M-5] **liststruct( )**.

3. Mata now engages in more thorough type checking, and produces better code, for those who explicitly declare arguments and variables.

4. Mata inherits all of Stata's formats and functions for dealing with the new date/time variables and values; see [M-5] **date( )** and [M-5] **fmtwidth( )**.

5. New functions `inbase()` and `frombase()` perform base conversions; see [M-5] **inbase( )**.

6. New function `floatround()` returns values rounded to float precision. This is Mata's equivalent of Stata's `float()` function. See [M-5] **floatround( )**.

7. New function `nameexternal()` returns the name of an external; see [M-5] **findexternal( )**.

8. Concerning matrix manipulation,

   a. Matrix multiplication is now faster when one of the matrices contains many zeros, as is function `cross()`.

   b. Appending rows or columns to a matrix using , and \ is now faster.

   c. New function `_diag()` replaces the principal diagonal of a matrix with a specified vector or scalar; see [M-5] **_diag( )**.

   d. New functions `select()` and `st_select()` select rows or columns of a matrix on the basis of a criterion; see [M-5] **select( )**.

   e. Existing functions `rowsum()`, `colsum()`, `sum()`, `quadrowsum()`, `quadcolsum()`, and `quadsum()` now allow an optional second argument that determines how missing values are handled; see [M-5] **sum( )**.

   f. New functions `runningsum()`, `quadrunningsum()`, `_runningsum()`, and `_quadrunningsum()` return the running sum of a vector; see [M-5] **runningsum( )**.

   g. New functions `minindex()` and `maxindex()` return the indices of minimums and maximums (including tied values) of a vector; see [M-5] **minindex( )**.

9. Concerning statistics,

   a. New Mata function `optimize()` performs minimization and maximization. You can code just the function, the function and its first derivatives, or the function and its first and second derivatives. Optimization techniques include Newton–Raphson, Davidon–Fletcher–Powell, Broyden–Fletcher–Goldfarb–Shanno, Berndt–Hall–Hall–Hausman, and the simplex method Nelder–Mead. See [M-5] **optimize( )**.

   b. New function `cvpermute()` forms permutations; see [M-5] **cvpermute( )**.

   c. New function `ghk()` provides the Geweke–Hajivassiliou–Keane multivariate normal simulator; see [M-5] **ghk( )**. New function `ghkfast()` is faster but a little more difficult to use; see [M-5] **ghkfast( )**.

   d. New functions `halton()`, `_halton()`, and `ghalton()` compute Halton and Hammersley sets; see [M-5] **halton( )**.

   e. The new density and probability functions available in Stata are also available in Mata, including `binomial()`, `binomialtail()`, `gammaptail()`, `invgammaptail()`, `invbinomialtail()`, `ibetatail()`, `invibetatail()`, `lnnormal()`, and `lnnormalden()`; see [M-5] **normal( )**.

Also, as in Stata, convergence and accuracy of many of the cumulatives, reverse cumulatives, and density functions have been improved.

    f. Existing Mata functions `mean()`, `variance()`, `quadvariance()`, `meanvariance()`, `quad-meanvariance()`, `correlation()`, and `quadcorrelation()` now make the weight argument optional. If not specified, unweighted estimates are returned. See [M-5] **mean( )**.

10. Concerning string processing,

    a. New function `stritrim()` replaces multiple, consecutive internal spaces with one space; see [M-5] **strtrim( )**.

    b. New functions `strtoreal()` and `_strtoreal()` convert strings to numeric values; see [M-5] **strtoreal( )**.

    c. New function `_substr()` substitutes a substring into an existing string; see [M-5] **_substr( )**.

    d. New function `invtokens()` is the inverse of the existing function `tokens()`; see [M-5] **invtokens( )**.

    e. New function `tokenget()` performs advanced parsing; see [M-5] **tokenget( )**.

11. Concerning I/O,

    a. New function `adosubdir()` returns the subdirectory in which Stata would search for a file; see [M-5] **adosubdir( )**. New function `pathsearchlist(`*fn*`)` returns a row vector of full paths/filenames specifying all the locations, in order, where Stata would look for the specified *fn* along the official Stata ado-path; see [M-5] **pathjoin( )**.

    b. New function `byteorder()` returns 1 if the computer is HILO and returns 2 if the computer is LOHI; see [M-5] **byteorder( )**.

    c. New functions `bufput()` and `bufget()` copy elements into and out of buffers; see [M-5] **bufio( )**.

    d. Existing function `cat()` now allows optional second and third arguments that specify the beginning and ending lines of the file to read; see [M-5] **cat( )**.

12. New functions `ferrortext()` and `freturncode()` obtain error messages and return codes following an I/O error; see [M-5] **ferrortext( )**.

13. Concerning the Stata interface,

    a. New function `stataversion()` returns the version of Stata that you are running, and new function `statasetversion()` allows setting it. See [M-5] **stataversion( )**.

    b. New function `setmore()` allows turning `more` on and off. New function `setmoreonexit()` allows restoring `more` to its original setting when execution ends. See [M-5] **more( )**.

Also See

[M-1] **first** — Introduction and first session

[M-6] **glossary** — Mata glossary of common terms

[M-1] Introduction and advice

Title

[**M-1**] **intro** — Introduction and advice

Contents

[M-1] Entry | Description

| Introductory material |
---
first | Introduction and first session
interactive | Using Mata interactively
ado | Using Mata with ado-files
help | Obtaining online help

| How Mata works & finding examples |
---
how | How Mata works
source | Viewing the source code

| Special topics |
---
returnedargs | Function arguments used to return results
naming | Advice on naming functions and variables
limits | Limits and memory utilization
tolerance | Use and specification of tolerances
permutation | An aside on permutation matrices and vectors
LAPACK | The LAPACK linear-algebra routines

Description

This section provides an introduction to Mata along with reference material common to all sections.

Remarks

The most important entry in this section is [M-1] **first**. Also see [M-6] **glossary**.

Also See

[M-0] **intro** — Introduction to the Mata manual

Title

[M-1] **ado** — Using Mata with ado-files

Description

This section provides advice to ado-file programmers on how to use Mata effectively and efficiently.

Remarks

For those interested in extending Stata by adding new commands, Mata is not a replacement for ado-files. Rather, the appropriate way to use Mata is to create subroutines used by your ado-files. Below we discuss how to do that under the following headings:

> *A first example*
> *Where to store the Mata functions*
> *Passing arguments to Mata functions*
> *Returning results to ado-code*
> *Advice: Use of matastrict*
> *Advice: Some useful Mata functions*

A first example

We will pretend that Stata cannot produce sums and that we want to write a new command for Stata that will report the sum of one variable. Here is our first take on how we might do this:

──────────────────────────────────── top of `varsum.ado` ────────

```
program varsum
    version 10
    syntax varname [if] [in]
    marksample touse
    mata: calcsum("`varlist'", "`touse'")
    display as txt "  sum = " as res r(sum)
end
version 10
mata:
void calcsum(string scalar varname, string scalar touse)
{
    real colvector  x

    st_view(x, ., varname, touse)
    st_numscalar("r(sum)", colsum(x))
}
end
```

──────────────────────────────────── end of `varsum.ado` ────────

Running this program from Stata results in the following output:

```
. varsum mpg
    sum = 1576
```

Note the following:

1. The ado-file has both ado-code and Mata code in it.

10

2. The ado-code handled all issues of parsing and identifying the subsample of the data to be used.

3. The ado-code called a Mata function to perform the calculation.

4. The Mata function received as arguments the names of two variables in the Stata dataset: the variable on which the calculation was to be made and the variable that identified the subsample of the data to be used.

5. The Mata function returned the result in $r()$, from where the ado-code could access it.

The outline that we showed above is a good one, although we will show you alternatives that, for some problems, are better.

Where to store the Mata functions

Our ado-file included a Mata function. You have three choices of where to put the Mata function:

1. You can put the code for the Mata function in the ado-file, as we did. To work, your .ado file must be placed where Stata can find it.

2. You can omit code for the Mata function from the ado-file, compile the Mata function separately, and store the compiled code (the object code) in a separate file called a .mo file. You place both your .ado and .mo files where Stata can find them.

3. You can omit the code for the Mata function from the ado-file, compile the Mata function separately, and store the compiled code in a .mlib library. Here both your .ado file and your .mlib library will need to be placed where Stata can find them.

We will show you how to do each of these alternatives, but before we do, let's consider the advantages and disadvantages of each:

1. Putting your Mata source code right in the ado-file is easiest, and it sure is convenient. The disadvantage is that Mata must compile the source code into object code, and that will slow execution. The cost is small because it occurs infrequently; Mata compiles the code when the ado-file is loaded and, as long as the ado-file is not dropped from memory, Stata and Mata will use the same compiled code over and over again.

2. Saving your Mata code as .mo files saves Mata from having to compile them at all. The source code is compiled only once—at the time you create the .mo file—and thereafter, it is the already-compiled copy that Stata and Mata will use.

   There is a second advantage: rather than the Mata functions (calcsum() here) being private to the ado-file, you will be able to use calcsum() in your other ado-files. calcsum() will become a utility always available to you. Perhaps calcsum() is not deserving of this honor.

3. Saving your Mata code in a .mlib library has the same advantages and disadvantages as (2); the only difference is that we save the precompiled code in a different way. The .mo/.mlib choice is of more concern to those who intend to distribute their ado-file to others. .mlib libraries allow you to place many Mata functions (subroutines for your ado-files) into one file, and so it is easier to distribute.

Let's restructure our ado-file to save calcsum() in a .mo file. First, we simply remove calcsum() from our ado-file, so it now reads

───────────────────────────────────── top of varsum.ado ─────────

```
program varsum
    version 10
    syntax varname [if] [in]
    marksample touse
    mata: calcsum("'varlist'", "'touse'")
    display as txt "  sum = " as res r(sum)
end
```

───────────────────────────────────── end of varsum.ado ─────────

Next, we enter Mata, enter our calcsum() program, and save the object code:

```
: void calcsum(string scalar varname, string scalar touse)
> {
>         real colvector  x
>
>         st_view(x, ., varname, touse)
>         st_numscalar("r(sum)", colsum(x))
> }
: mata mosave calcsum(), dir(PERSONAL)
```

This step we do only once. The mata mosave command created file calcsum.mo stored in our PERSONAL sysdir directory; see [M-3] **mata mosave** and [P] **sysdir** for more details. We put the calcsum.mo file in our PERSONAL directory so that Stata and Mata would be able to find it, just as you probably did with the varsum.ado ado-file.

Typing in the calcsum() program interactively is too prone to error, so instead what we can do is create a do-file to perform the step and then run the do-file once:

───────────────────────────────────── top of varsum.do ─────────

```
version 10
mata:
void calcsum(string scalar varname, string scalar touse)
{
    real colvector  x
    st_view(x, ., varname, touse)
    st_numscalar("r(sum)", colsum(x))
}
mata mosave calcsum(), dir(PERSONAL) replace
end
```

───────────────────────────────────── end of varsum.do ─────────

We save the do-file someplace safe in case we should need to modify our code later, either to fix bugs or to add features. For the same reason, we added a replace option on the command that creates the .mo file, so we can run the do-file over and over.

To follow the third organization—putting our calcsum() subroutine in a library—is just a matter of modifying varsum.do to do mata mlib add rather than mata mosave. See [M-3] **mata mlib**; there are important details having to do with how you will manage all the different functions you will put in the library, but that has nothing to do with our problem here.

Where and how you store your Mata subroutines has nothing to do with what your Mata subroutines do or how you use them.

Passing arguments to Mata functions

In designing a subroutine, you have two paramount interests: how you get data into your subroutine and how you get results back. You get data in by the values you pass as the arguments. For `calcsum()`, we called the subroutine by coding

```
mata: calcsum("'varlist'", "'touse'")
```

After macro expansion, that line turned into something like

```
mata: calcsum("mpg", "__0001dc")
```

The `__0001dc` variable is a temporary variable created by the `marksample` command earlier in our ado-file. `mpg` was the variable specified by the user. After expansion, the arguments were nothing more than strings, and those strings were passed to `calcsum()`.

Macro substitution is the most common way values are passed to Mata subroutines. If we had a Mata function `add(a, b)`, which could add numbers, we might code in our ado-file

```
mata: add('x', 'y')
```

and, if macro 'x' contained 2 and macro 'y' contained 3, Mata would see

```
mata: add(2, 3)
```

and values 2 and 3 would be passed to the subroutine.

When you think about writing your Mata subroutine, the arguments your ado-file will find convenient to pass and Mata will make convenient to use are

1. numbers, which Mata calls real scalars, such as 2 and 3 ('x' and 'y'), and

2. names of variables, macros, scalars, matrices, etc., which Mata calls string scalars, such as "mpg" and "__0001dc" ("'varlist'" and "'touse'").

To receive arguments of type (1), you code `real scalar` as the type of the argument in the function declaration and then use the real scalar variable in your Mata code.

To receive arguments of type (2), you code `string scalar` as the type of the argument in the function declaration, and then you use one of the Stata interface functions (in [M-4] **stata**) to go from the name to the contents. If you receive a variable name, you will especially want to read about the functions `st_data()` and `st_view()` (see [M-5] **st_data( )** and [M-5] **st_view( )**), although there are many other utilities for dealing with variable names. If you are dealing with local or global macros, scalars, or matrices, you will want to see [M-5] **st_local( )**, [M-5] **st_global( )**, [M-5] **st_numscalar( )**, and [M-5] **st_matrix( )**.

Returning results to ado-code

You have many more choices on how to return results from your Mata function to the calling ado-code.

1. You can return results in `r()`—as we did in our example—or in `e()` or in `s()`.

2. You can return results in macros, scalars, matrices, etc., whose names are passed to your Mata subroutine as arguments.

3. You can highhandedly reach back into the calling ado-file and return results in macros, scalars, matrices, etc., whose names are of your devising.

In all cases, see [M-5] **st_global( )**. st_global() is probably not the function you will use, but there is a wonderfully useful table in the *Remarks* section that will tell you exactly which function to use.

Also see all other Stata interface functions in [M-4] **stata**.

If you want to modify the Stata dataset in memory, see [M-5] **st_store( )** and [M-5] **st_view( )**.

Advice: Use of matastrict

The setting matastrict determines whether declarations can be omitted (see [M-2] **declarations**); by default, you may. That is, matastrict is set off, but you can turn it on by typing mata set matastrict on; see [M-3] **mata set**. Some users do, and some users do not.

So now, consider what happens when you include Mata source code directly in the ado-file. When the ado-file is loaded, is matastrict set on, or is it set off? The answer is that it is off, because when you include the Mata source code inside an ado-file, matastrict is temporarily switched off when the ado-file is loaded even if the user running the ado-file has previously set it on.

For example, varsum.ado could read

――――――――――――――――――――――――――― top of `varsum.ado` ―――――――

```
program varsum
    version 10
    syntax varname [if] [in]
    marksample touse
    mata: calcsum("`varlist'", "`touse'")
    display as txt "  sum = " as res r(sum)
end

version 10
mata:
void calcsum(varname, touse)
{                                        // (note absence of declarations)
    st_view(x, ., varname, touse)
    st_numscalar("r(sum)", colsum(x))
}
end
```

―――――――――――――――――――――――――― end of `varsum.ado` ―――――――

and it will work even when run by users who have set matastrict on.

Similarly, in an ado-file, you can set matastrict on and that will not affect the setting after the ado-file is loaded, so varsum.ado could read

─────────────────────────────────────── top of `varsum.ado` ───────────

```
program varsum
    version 10
    syntax varname [if] [in]
    marksample touse
    mata: calcsum("`varlist'", "`touse'")
    display as txt "  sum = " as res r(sum)
end
version 10
mata:
mata set matastrict on
void calcsum(string scalar varname, string scalar touse)
{
    real colvector  x
    st_view(x, ., varname, touse)
    st_numscalar("r(sum)", colsum(x))
}
end
```

───────────────────────────────── end of `varsum.ado` ───────────

and not only will it work, but running `varsum` will not change the user's `matastrict` setting.

This preserving and restoring of `matastrict` is something that is done only for ado-files when they are loaded.

Advice: Some useful Mata functions

In the `calcsum()` subroutine, we used the `colsum()` function—see [M-5] **sum( )**—to obtain the sum:

```
void calcsum(string scalar varname, string scalar touse)
{
    real colvector  x
    st_view(x, ., varname, touse)
    st_numscalar("r(sum)", colsum(x))
}
```

We could instead have coded

```
void calcsum(string scalar varname, string scalar touse)
{
    real colvector  x
    real scalar     i, sum
    st_view(x, ., varname, touse)
    sum = 0
    for (i=1; i<=rows(x); i++) sum = sum + x[i]
    st_numscalar("r(sum)", sum)
}
```

The first way is preferred. Rather than construct explicit loops, it is better to call functions that calculate the desired result when such functions exist. Unlike ado-code, however, when such functions do not exist, you can code explicit loops and still obtain good performance.

Another set of functions we recommend are documented in [M-5] **cross( )**, [M-5] **crossdev( )**, and [M-5] **quadcross( )**.

cross() makes calculations of the form

$$X'X$$
$$X'Z$$
$$X'\text{diag}(w)X$$
$$X'\text{diag}(w)Z$$

crossdev() makes calculations of the form

$$(X{:}{-}x)'(X{:}{-}x)$$
$$(X{:}{-}x)'(Z{:}{-}z)$$
$$(X{:}{-}x)'\text{diag}(w)(X{:}{-}x)$$
$$(X{:}{-}x)'\text{diag}(w)(Z{:}{-}z)$$

Both these functions could easily escape your attention because the matrix expressions themselves are so easily written in Mata. The functions, however, are quicker, use less memory, and sometimes are more accurate. Also, quad-precision versions of the functions exist; [M-5] **quadcross( )**.

Also See

[M-2] **version** — Version control

[M-1] **intro** — Introduction and advice

Title

[M-1] **first** — Introduction and first session

Description

Mata is a component of Stata. It is a matrix programming language which can be used interactively or as an extension for do-files and ado-files. Thus

1. Mata can be used by users who want to think in matrix terms and perform (not necessarily simple) matrix calculations interactively, and

2. Mata can be used by advanced Stata programmers who want to add features to Stata.

Mata has something for everybody.

Primary features of Mata are that it is fast and that it is C-like.

Remarks

This introduction is presented under the following headings:

> *Invoking Mata*
> *Using Mata*
> *Making mistakes: Interpreting error messages*
> *Working with real numbers, complex numbers, and strings*
> *Working with scalars, vectors, and matrices*
> *Working with functions*
> *Distinguishing real and complex values*
> *Working with matrix and scalar functions*
> *Performing element-by-element calculations: Colon operators*
> *Writing programs*
> *More functions*
> *Mata environment commands*
> *Exiting Mata*

If you are reading the entries in the order suggested in [M-0] **intro**, see [M-1] **interactive** next.

Invoking Mata

To enter Mata, type `mata` at Stata's dot prompt and press enter; to exit Mata, type `end` at Mata's colon prompt:

```
. mata
———————— mata (type end to exit) ——          ← type mata to enter Mata
: 2 + 2                                        ← type Mata statements at the colon prompt
  4

: end                                          ← type end to return to Stata

. _                                            ← you are back to Stata
```

17

Using Mata

When you type a statement into Mata, Mata compiles what you typed and, if it compiled without error, executes it:

```
: 2 + 2
  4

:  _
```

We typed 2 + 2, a particular example from the general class of expressions. Mata responded with 4, the evaluation of the expression.

Often what you type are expressions, although you will probably choose more complicated examples. When an expression is not assigned to a variable, the result of the expression is displayed. Assignment is performed by the = operator:

```
: x = 2 + 2

: x
  4

:  _
```

When we type "x = 2 + 2", the expression is evaluated and stored in the variable we just named x. The result is not displayed. We can look at the contents of x, however, simply by typing "x". From Mata's perspective, x is not only a variable but also an expression, albeit a rather simple one. Just as 2 + 2 says to load 2, load another 2, and add them, the expression x says to load x and stop there.

As an aside, Mata distinguishes uppercase and lowercase. X is not the same as x:

```
: X = 2 + 3

: x
  4

: X
  5

:  _
```

Making mistakes: Interpreting error messages

If you make a mistake, Mata complains, and then you continue on your way. For instance,

```
: 2,,3
invalid expression
r(3000);

:  _
```

2,,3 makes no sense to Mata, so Mata complained. This is an example of what is called a compile-time error; Mata could not make sense out of what we typed.

The other kind of error is called a run-time error. For example, we have no variable called y. Let us ask Mata to show us the contents of y:

```
: y
                    <istmt>:  3499  y not found
r(3499);

:  _
```

Here what we typed made perfect sense—show me y—but y has never been defined. This ugly message is called a run-time error message—see [M-2] **errors** for a complete description—but all that's important is to understand the difference between

```
invalid expression
```

and

```
<istmt>:   3499  y not found
```

The run-time message is prefixed by an identity (<istmt> here) and a number (3499 here). Mata is telling us, "I was executing your *istmt* [that's what everything you type is called] and I got error 3499, the details of which are that I was unable to find y."

The compile-time error message is of a simpler form: invalid expression. When you get such unprefixed error messages, that means Mata could not understand what you typed. When you get the more complicated error message, that means Mata understood what you typed, but there was a problem performing your request.

Another way to tell the difference between compile-time errors and run-time errors is to look at the return code. Compile-time errors have a return code of 3000:

```
: 2,,3
invalid expression
r(3000);
```

Run-time errors have a return code that might be in the 3000s, but is never 3000 exactly:

```
: y
                <istmt>:   3499  y not found
r(3499);
```

Whether the error is compile-time or run-time, once the error message is issued, Mata is ready to continue just as if the error never happened.

Working with real numbers, complex numbers, and strings

As we have seen, Mata works with real numbers:

```
: 2 + 3
  5
```

Mata also understands complex numbers; you write the imaginary part by suffixing a lowercase i:

```
: 1+2i + 4-1i
  5 + 1i
```

For imaginary numbers, you can omit the real part:

```
: 1+2i - 2i
  1
```

Whether a number is real or complex, you can use the same computer notation for the imaginary part as you would for the real part:

```
: 2.5e+3i
  2500i
: 1.25e+2+2.5e+3i              /* i.e., 1.25+e02 + 2.5e+03i */
  125 + 2500i
```

We purposely wrote the last example in nearly unreadable form just to emphasize that Mata could interpret it.

Mata also understands strings, which you write enclosed in double quotes:

```
: "Alpha" + "Beta"
AlphaBeta
```

Just like Stata, Mata understands simple and compound double quotes:

```
: '"Alpha"' + '"Beta"'
AlphaBeta
```

You can add complex and reals,

```
: 1+2i + 3
4+2i
```

but you may not add reals or complex to strings:

```
: 2 + "alpha"
                    <istmt>:   3250   type mismatch
r(3250);
```

We got a run-time error. Mata understood 2 + "alpha" all right; it just could not perform our request.

Working with scalars, vectors, and matrices

In addition to understanding scalars—be they real, complex, or string—Mata understands vectors and matrices of real, complex, and string elements:

```
: x = (1, 2)
: x
         1    2

    1 |  1    2 |
```

x now contains the row vector (1, 2). We can add vectors:

```
: x + (3, 4)
         1    2

    1 |  4    6 |
```

The "," is the column-join operator; things like (1, 2) are expressions, just as (1 + 2) is an expression:

```
: y = (3, 4)
: z = (x, y)
: z
         1    2    3    4

    1 |  1    2    3    4 |
```

In the above, we could have dispensed with the parentheses and typed "y = 3, 4" followed by "z = x, y", just as we could using the + operator, although most people find vectors more readable when enclosed in parentheses.

\ is the row-join operator:

```
: a = (1 \ 2)
: a
        1

  1 │  1
  2 │  2

:
: b = (3 \ 4)
: c = (a \ b)
: c
        1

  1 │  1
  2 │  2
  3 │  3
  4 │  4
```

Using the column-join and row-join operators, we can enter matrices:

```
: A = (1, 2 \ 3, 4)
: A
        1    2

  1 │  1    2
  2 │  3    4
```

The use of these operators is not limited to scalars. Remember, x is the row vector (1, 2), y is the
row vector (3, 4), a is the column vector (1 \ 2), and b is the column vector (3 \ 4). Therefore,

```
: x \ y
        1    2

  1 │  1    2
  2 │  3    4

: a, b
        1    2

  1 │  1    3
  2 │  2    4
```

But if we try something nonsensical, we get an error:

```
: a, x
                <istmt>:  3200  conformability error
```

We create complex vectors and matrices just as we create real ones, the only difference being that
their elements are complex:

```
: Z = (1 + 1i, 2 + 3i \ 3 - 2i, -1 - 1i)
: Z
                1          2

  1 │   1 + 1i     2 + 3i
  2 │   3 - 2i    -1 - 1i
```

Similarly, we can create string vectors and matrices, which are vectors and matrices with string elements:

```
: S = ("1st element", "2nd element" \ "another row", "last element")
: S
                      1                2

  1       1st element    2nd element
  2       another row    last element
```

For strings, the individual elements can be up to 2,147,483,647 characters long.

Working with functions

Mata's expressions also include functions:

```
: sqrt(4)
  2
: sqrt(-4)
  .
```

When we ask for the square root of −4, Mata replies ".". Further, . can be stored just like any other number:

```
: findout = sqrt(-4)
: findout
  .
```

"." means missing, that there is no answer to our calculation. Taking the square root of a negative number is not an error; it merely produces missing. To Mata, missing is a number like any other number, and the rules for all the operators have been generalized to understand missing. For instance, the addition rule is generalized such that missing plus anything is missing:

```
: 2 + .
  .
```

Still, it should surprise you that Mata produced missing for the sqrt(-4). We said that Mata understands complex numbers, so should not the answer be 2i? The answer is that is should be if you are working on the complex plane, but otherwise, missing is probably a better answer. Mata attempts to intuit the kind of answer you want by context, and in particular, uses inheritance rules. If you ask for the square root of a real number, you get a real number back. If you ask for the square root of a complex number, you get a complex number back:

```
: sqrt(-4 + 0i)
  2i
```

Here complex means multipart: −4 + 0i is a complex number; it merely happens to have 0 imaginary part. Thus:

```
: areal = -4
: acomplex = -4 + 0i
: sqrt(areal)
  .
: sqrt(acomplex)
  2i
```

If you ever have a real scalar, vector, or matrix, and want to make it complex, use the C() function, which means "convert to complex":

```
: sqrt(C(areal))
  2i
```

C() is documented in [M-5] **C()**. C() allows one or two arguments. With one argument, it casts to complex. With two arguments, it makes a complex out of the two real arguments. Thus you could type

```
: sqrt(-4 + 2i)
  .485868272 + 2.05817103i
```

or you could type

```
: sqrt(C(-4, 2))
  .485868272 + 2.05817103i
```

By the way, used with one argument, C() also allows complex, and then it does nothing:

```
: sqrt(C(acomplex))
  2i
```

Distinguishing real and complex values

It is virtually impossible to tell the difference between a real value and a complex value with zero imaginary part:

```
: areal = -4
: acomplex = -4 + 0i
: areal
  -4
: acomplex
  -4
```

Yet, as we have seen, the difference is important: sqrt(areal) is missing, sqrt(acomplex) is 2i. One solution is the eltype() function:

```
: eltype(areal)
  real
: eltype(acomplex)
  complex
```

eltype() can also be used with strings,

```
: astring = "hello"
: eltype(astring)
  string
```

but this is useful mostly in programming contexts.

Working with matrix and scalar functions

Some functions are matrix functions: they require a matrix and return a matrix. Mata's $invsym(X)$ is an example of such a function. It returns the matrix that is the inverse of symmetric, real matrix X:

```
: X = (76, 53, 48 \ 53, 88, 46 \ 48, 46, 63)

: Xi = invsym(X)

: Xi
[symmetric]
                  1                2                3

    1    .0298458083
    2   -.0098470272     .0216268926
    3   -.0155497706    -.0082885675     .0337724301

: Xi * X
                  1                2                3

    1                1    -8.67362e-17    -8.50015e-17
    2    -1.38778e-16                1    -1.02349e-16
    3                0     1.11022e-16                1
```

The last matrix, $Xi * X$, differs just a little from the identity matrix because of unavoidable computational roundoff error.

Other functions are, mathematically speaking, scalar functions. $sqrt()$ is an example in that it makes no sense to speak of $sqrt(X)$. (That is, it makes no sense to speak of $sqrt(X)$ unless we were speaking of the Cholesky square-root decomposition. Mata has such a matrix function; see [M-5] **cholesky( )**.)

When a function is, mathematically speaking, a scalar function, the corresponding Mata function will usually allow vector and matrix arguments and, then, the Mata function makes the calculation on each element individually:

```
: M = (1, 2 \ 3, 4 \ 5, 6)

: M
        1    2

    1    1    2
    2    3    4
    3    5    6

:
: S = sqrt(M)

: S
                  1                2

    1                1     1.414213562
    2    1.732050808                2
    3    2.236067977     2.449489743

:
: S[1,2]*S[1,2]
  2

: S[2,1]*S[2,1]
  3
```

When a function returns a result calculated in this way, it is said to return an element-by-element result.

Performing element-by-element calculations: Colon operators

Mata's operators, such as + (addition) and * (multiplication), work as you would expect. In particular, * performs matrix multiplication:

```
: A = (1, 2 \ 3, 4)
: B = (5, 6 \ 7, 8)
: A*B
         1     2

   1    19    22
   2    43    50
```

The first element of the result was calculated as $1 * 5 + 2 * 7 = 19$.

Sometimes, you really want the element-by-element result. When you do, place a colon in front of the operator: Mata's :* operator performs element-by-element multiplication:

```
: A:*B
         1     2

   1     5    12
   2    21    32
```

See [M-2] **op_colon** for more information.

Writing programs

Mata is a complete programming language; it will allow you to create your own functions:

```
: function add(a,b) return(a+b)
```

That single statement creates a new function, although perhaps you would prefer if we typed it as

```
: function add(a,b)
> {
>         return(a+b)
> }
```

because that makes it obvious that a program can contain many lines. In either case, once defined, we can use the function:

```
: add(1,2)
  3
: add(1+2i,4-1i)
  5 + 1i
: add( (1,2), (3,4) )
       1    2

   1   4    6
```

```
: add(x,y)
        1    2
    ┌──────────┐
  1 │  4    6  │
    └──────────┘

: add(A,A)
        1    2
    ┌──────────┐
  1 │  2    4  │
  2 │  6    8  │
    └──────────┘

:
: Z1 = (1+1i, 1+1i \ 2, 2i)
: Z2 = (1+2i, -3+3i \ 6i, -2+2i)
: add(Z1, Z2)
            1            2
    ┌────────────────────────┐
  1 │    2 + 3i    -2 + 4i    │
  2 │    2 + 6i    -2 + 4i    │
    └────────────────────────┘

:
: add("Alpha","Beta")
  AlphaBeta

:
: S1 = ("one", "two" \ "three", "four")
: S2 = ("abc", "def" \ "ghi", "jkl")
: add(S1, S2)
            1            2
    ┌────────────────────────┐
  1 │  oneabc       twodef    │
  2 │  threeghi     fourjkl   │
    └────────────────────────┘
```

Of course, our little function `add()` does not do anything that the + operator does not already do, but we could write a program that did do something different. The following program will allow us to make $n \times n$ identity matrices:

```
: real matrix id(real scalar n)
> {
>        real scalar i
>        real matrix res
>
>        res = J(n, n, 0)
>            for (i=1; i<=n; i++) {
>                    res[i,i] = 1
>            }
>            return(res)
> }

:
: I3 = id(3)

: I3
[symmetric]
        1    2    3
    ┌──────────────┐
  1 │  1           │
  2 │  0    1      │
  3 │  0    0    1 │
    └──────────────┘
```

The function J() in the program line res = J(n, n, 0) is a Mata built-in function that returns an $n \times n$ matrix containing 0s ($J(r, c, val)$ returns an $r \times c$ matrix, the elements of which are all equal to *val*); see [M-5] **J( )**.

for (i=1; i<=n; i++) says that starting with i=1 and so long as i<=n do what is inside the braces (set res[i,i] equal to 1) and then (we are back to the for part again), increment i.

The final line—return(res)—says to return the matrix we have just created.

Actually, just as with add(), we do not need id() because Mata has a built-in function I(n) that makes identity matrices, but it is interesting to see how the problem could be programmed.

More functions

Mata has many functions already and much of this manual concerns documenting what those functions do; see [M-4] **intro**. But right now, what is important is that many of the functions are themselves written in Mata!

One of those functions is pi(); it takes no arguments and returns the value of *pi*. The code for it reads

```
real scalar pi() return(3.141592653589793238462643)
```

There is no reason to type the above function because it is already included as part of Mata:

```
: pi()
  3.141592654
```

When Mata lists a result, it does not show as many digits, but we could ask to see more:

```
: printf("%17.0g", pi())
  3.14159265358979
```

Other Mata functions include the hyperbolic functions $\sinh(u)$, $\cosh(u)$, etc. The code for $\sinh(u)$, $\cosh(u)$, and $\tanh(u)$ reads

```
numeric matrix sinh(numeric matrix u) return((exp(u)-exp(-u)):/2)

numeric matrix cosh(numeric matrix u) return((exp(u)+exp(-u)):/2)

numeric matrix tanh(numeric matrix u)
{
        numeric matrix   eu, emu

        eu = exp(u)
        emu = exp(-u)
        return( (eu-emu):/(eu+emu) )
}
```

See for yourself: at the Stata dot prompt (not the Mata colon prompt), type

```
. viewsource sinh.mata

. viewsource cosh.mata

. viewsource tanh.mata
```

When the code for a function was written in Mata (as opposed to having been written in C), viewsource can show you the code; see [M-1] **source**.

Returning to the functions themselves,

```
numeric matrix sinh(numeric matrix u) return((exp(u)-exp(-u)):/2)

numeric matrix cosh(numeric matrix u) return((exp(u)+exp(-u)):/2)

numeric matrix tanh(numeric matrix u)
{
    numeric matrix  eu, emu

    eu = exp(u)
    emu = exp(-u)
    return( (eu-emu):/(eu+emu) )
}
```

this is the first time we have seen the word `numeric`: it means real or complex. Built-in (previously written) function `exp()` works like `sqrt()` in that it allows a real or complex argument and correspondingly returns a real or complex result. Said in Mata jargon: `exp()` allows a `numeric` argument and correspondingly returns a `numeric` result. `sinh()`, `cosh()`, and `tanh()` will also work like `sqrt()` and `exp()`.

Another characteristic `sinh()`, `cosh()`, and `tanh()` share with `sqrt()` and `exp()` is element-by-element operation. `sinh()`, `cosh()`, and `tanh()` are element-by-element because `exp()` is element-by-element and because we were careful to use the `:/` (element-by-element) divide operator.

In any case, there is no need to type the above functions because they are already part of Mata. You could learn more about them by seeing their manual entry, [M-5] **sin( )**.

At the other extreme, Mata functions can become long. Here is Mata's function to solve $AX = B$ for X when A is lower triangular, placing the result X back into A:

```
real scalar _solvelower(
        numeric matrix A, numeric matrix b,
        |real scalar usertol, numeric scalar userd)
{
    real scalar          tol, rank, a_t, b_t, d_t
    real scalar          n, m, i, im1, complex_case
    numeric rowvector    sum
    numeric scalar       zero, d

    d  = userd

    if ((n=rows(A))!=cols(A)) _error(3205)
    if (n != rows(b))        _error(3200)
    if (isview(b))           _error(3104)
    m = cols(b)
    rank = n

    a_t = iscomplex(A)
    b_t = iscomplex(b)
    d_t = d<. ? iscomplex(d) : 0

    complex_case = a_t | b_t | d_t
```

```
    if (complex_case) {
        if (!a_t) A = C(A)
        if (!b_t) b = C(b)
        if (d<. & !d_t) d = C(d)
        zero = 0i
    }
    else zero = 0

    if (n==0 | m==0) return(0)

    tol = solve_tol(A, usertol)

    if (abs(d) >=. ) {
        if (abs(d=A[1,1])<=tol) {
            b[1,.] = J(1, m, zero)
            --rank
        }
        else {
            b[1,.] = b[1,.] :/ d
            if (missing(d)) rank = .
        }

        for (i=2; i<=n; i++) {
            im1 = i - 1
            sum = A[|i,1\i,im1|] * b[|1,1\im1,m|]
            if (abs(d=A[i,i])<=tol) {
                b[i,.] = J(1, m, zero)
                --rank
            }
            else {
                b[i,.] = (b[i,.]-sum) :/ d
                if (missing(d)) rank = .
            }
        }
    }
    else {
        if (abs(d)<=tol) {
            rank = 0
            b = J(rows(b), cols(b), zero)
        }
        else {
            b[1,.] = b[1,.] :/ d

            for (i=2; i<=n; i++) {
                im1 = i - 1
                sum = A[|i,1\i,im1|] * b[|1,1\im1,m|]
                b[i,.] = (b[i,.]-sum) :/ d
            }
        }

    }
    return(rank)
}
```

If the function were not already part of Mata and you wanted to use it, you could type it into a do-file or onto the end of an ado-file (especially good if you just want to use _solvelower() as a subroutine). In those cases, do not forget to enter and exit Mata:

———————————————————————————————————— top of ado-file ————————

```
program mycommand
      . . .
      ado-file code appears here
      . . .
end
mata:
      _solvelower() code appears here
end
```

———————————————————————————————————— end of ado-file ————————

Sharp-eyed readers will notice that we put a colon on the end of the Mata command. That's a detail, and why we did that is explained in [M-3] **mata**.

In addition to loading functions by putting their code in do- and ado-files, you can also save the compiled versions of functions in .mo files (see [M-3] **mata mosave**) or into .mlib Mata libraries (see [M-3] **mata mlib**).

For _solvelower(), it has already been saved into a library, namely, Mata's official library, so you need not do any of this.

Mata environment commands

When you are using Mata, there is a set of commands that will tell you about and manipulate Mata's environment.

The most useful such command is mata describe:

```
: mata describe
      # bytes    type                     name and extent
            76   transmorphic matrix      add()
           200   real matrix              id()
            32   real matrix              A[2,2]
            32   real matrix              B[2,2]
            72   real matrix              I3[3,3]
            48   real matrix              M[3,2]
            48   real matrix              S[3,2]
            47   string matrix            S1[2,2]
            44   string matrix            S2[2,2]
            72   real matrix              X[3,3]
            72   real matrix              Xi[3,3]
            64   complex matrix           Z[2,2]
            64   complex matrix           Z1[2,2]
            64   complex matrix           Z2[2,2]
            16   real colvector           a[2]
            16   complex scalar           acomplex
             8   real scalar              areal
            16   real colvector           b[2]
            32   real colvector           c[4]
             8   real scalar              findout
            16   real rowvector           x[2]
            16   real rowvector           y[2]
            32   real rowvector           z[4]

: _
```

Another useful command is `mata clear`, which will clear Mata without disturbing Stata:

```
: mata clear
: mata describe
      # bytes   type                    name and extent
```

There are other useful `mata` commands; see [M-3] **intro**. Do not confuse this command `mata`, which you type at Mata's colon prompt, with Stata's command `mata`, which you type at Stata's dot prompt and which invokes Mata.

Exiting Mata

When you are done using Mata, type `end` to Mata's colon prompt:

```
: end

. _
```

Exiting Mata does not clear it:

```
. mata
                                   ───── mata (type end to exit) ─────
: x = 2
: y = (3 + 2i)
: function add(a,b) return(a+b)
: end

. ...
. mata
                                   ───── mata (type end to exit) ─────
: mata describe
      # bytes   type                    name and extent

         38     transmorphic matrix     add()
          8     real scalar             x
         16     complex scalar          y

: end
```

Exiting Stata clears Mata, as does Stata's `clear mata` command; see [D] **clear**.

Also See

[M-1] **intro** — Introduction and advice

Title

> **[M-1] help** — Obtaining online help

Syntax

> help m# *entryname*
>
> help mata *functionname*()

The help command may be issued at either Stata's dot prompt or Mata's colon prompt.

Description

Help for Mata is available online in Stata. This entry describes how to access it.

Remarks

To see this entry online, type

> . help m1 help

at Stata's dot prompt or Mata's colon prompt. You type that because this entry is [M-1] **help**.

To see the entry for function max(), for example, type

> . help mata max()

max() is documented in [M-5] **minmax( )**, but that will not matter; Stata will find the appropriate entry.

To enter the Mata help system from the top, from whence you can click your way to any section or function, type

> . help mata

Also See

[R] **help** — Display online help

[M-3] **mata help** — Obtain online help

[M-1] **intro** — Introduction and advice

Title

Description

Below we take away some of the mystery and show you how Mata works. Everyone, we suspect, will find this entertaining, and advanced users will find the description useful for predicting what Mata will do when faced with unusual situations.

Remarks

Remarks are presented under the following headings:

> *What happens when you define a program*
> *What happens when you work interactively*
> *What happens when you type a mata environment command*
> *Working with object code I: .mo files*
> *Working with object code II: .mlib libraries*
> *The Mata environment*

If you are reading the entries in the order suggested in [M-0] **intro**, browse [M-1] **intro** next for sections that interest you, and then see [M-2] **syntax**.

What happens when you define a program

Let's say you fire up Mata and type

```
: function tryit()
> {
>     real scalar i
>
>     for (i=1; i<=10; i++) i
> }
```

Mata compiles the program: it reads what you type and produces binary codes that tell Mata exactly what it is to do when the time comes to execute the program. In fact, given the above program, Mata produces the binary code

```
00b4 3608 4000 0000 0100 0000 2000 0000
0000 0000 ffff ffff 0300 0000 0000 0000
0100 7472 7969 7400 1700 0100 1f00 0700
0000 0800 0000 0200 0100 0800 2a00 0300
1e00 0300
```

which looks meaningless to you and me, but Mata knows exactly what to make of it. The compiled version of the program is called object code, and it is the object code, not the original source code, that Mata stores in memory. In fact, the original source is discarded once the object code has been stored.

It is this compilation step—the conversion of text into object code—that makes Mata able to execute programs so quickly.

Later, when the time comes to execute the program, Stata follows the instructions it has previously recorded:

```
: tryit()
  1
  2
  3
  4
  5
  6
  7
  8
  9
  10
```

What happens when you work interactively

Let's say you type

```
: x = 3
```

In the jargon of Mata, that is called an *istmt*—an interactive statement. Obviously, Mata stores 3 in x, but how?

Mata first compiles the single statement and stores the resulting object code under the name <istmt>. The result is much as if you had typed

```
: function <istmt>()
> {
>           x = 3
> }
```

except, of course, you could not define a program named <istmt> because the name is invalid. Mata has ways of getting around that.

At this point, Mata has discarded the source code x = 3 and has stored the corresponding object code. Next, Mata executes <istmt>. The result is much as if you had typed

```
: <istmt>()
```

That done, there is only one thing left to do, which is to discard the object code. The result is much as if you typed

```
: mata drop <istmt>()
```

So there you have it: you type

```
: x = 3
```

and Mata executes

```
: function <istmt>()
> {
>           x = 3
> }
: <istmt>()
: mata drop <istmt>()
```

❑ Technical Note

The above story is not exactly true because, as told, variable x would be local to function `<istmt>()` so, when `<istmt>()` concluded execution, variable x would be discarded. To prevent that from happening, Mata makes all variables defined by `<istmt>()` global. Thus you can type

```
: x = 3
```

followed by

```
: y = x + 2
```

and all works out just as you expect: y is set to 5.

❑

What happens when you type a mata environment command

When you are at a colon prompt and type something that begins with the word `mata`, such as

```
: mata describe
```

or

```
: mata clear
```

something completely different happens: Mata freezes itself and temporarily transfers control to a command interpreter like Stata's. The command interpreter accesses Mata's environment and reports on it or changes it. Once done, the interpreter returns to Mata, which comes back to life, and issues a new colon prompt:

```
: _
```

Once something is typed at the prompt, Mata will examine it to determine if it begins with the word `mata` (in which case the interpretive process repeats), or if it is the beginning of a function definition (in which case the program will be compiled but not executed), or anything else (in which case Mata will try to compile and execute it as an `<istmt>()`).

Working with object code I: .mo files

Everything hinges on the object code that Mata produces, and, if you wish, you can save the object code on disk. The advantage of doing this is that, at some future date, your program can be executed without compilation, which saves time. If you send the object code to others, they can use your program without ever seeing the source code behind it.

After you type, say,

```
: function tryit()
> {
>        real scalar i
>
>        for (i=1; i<=10; i++) i
> }
```

Mata has created the object code and discarded the source. If you now type

```
: mata mosave tryit()
```

the Mata interpreter will create file `tryit.mo` in the current directory; see [M-3] **mata mosave**. The new file will contain the object code.

At some future date, were you to type

```
: tryit()
```

without having first defined the program, Mata would look along the ado-path (see [P] **sysdir** and [U] **17.5 Where does Stata look for ado-files?**) for a file named tryit.mo. Finding the file, Mata loads it (so Mata now has the object code and executes it in the usual way).

Working with object code II: .mlib libraries

Object code can be saved in .mlib libraries (files) instead of .mo files. .mo files contain the object code for one program. .mlib files contain the object code for a group of files.

The first step is to choose a name (we will choose lmylib—library names are required to start with the lowercase letter *l*) and create an empty library of that name:

```
: mlib create lmylib
```

Once created, new functions can be added to the library:

```
: mlib add lmylib tryit()
```

New functions can be added at any time, either immediately after creation or later—even much later; see [M-3] **mata mlib**.

We mentioned that when Mata needs to execute a function that it does not find in memory, Mata looks for a .mo file of the same name. Before Mata does that, however, Mata thumbs through its libraries to see if it can find the function there.

The Mata environment

Certain settings of Mata affect how it behaves. You can see those settings by typing mata query at the Mata prompt:

```
: mata query

    Mata settings
        set matastrict      off
        set matalnum        off
        set mataoptimize    on
        set matafavor       space      may be space or speed
        set matacache       400            kilobytes
        set matalibs        lmatabase;lmataopt;lmataado
        set matamofirst     off

    : _
```

You can change these settings by using mata set; see [M-3] **mata set**. We recommend the default settings, except that we admit to being partial to mata set matastrict on.

Reference

Gould, W. 2006. Mata Matters: Precision. *Stata Journal* 6: 550–560.

Also See

[M-1] **intro** — Introduction and advice

Title

> **[M-1] interactive** — Using Mata interactively

Description

With Mata, you simply type matrix formulas to obtain the desired results. Below we provide guidelines when doing this with statistical formulas.

Remarks

You have data and statistical formulas that you wish to calculate, such as $b = (X'X)^{-1}X'y$. Perform the following nine steps:

1. Start in Stata. Load the data.

2. If you are doing time-series analysis, generate new variables containing any *op.varname* variables you need, such as `l.gnp` or `d.r`.

3. Create a constant variable (`. gen cons = 1`). In most statistical formulas, you will find it useful.

4. Drop variables that you will not need. This saves memory and makes some things easier because you can just refer to all the variables.

5. Drop observations with missing values. Mata understands missing values, but Mata is a matrix language, not a statistical system, so Mata does not always ignore observations with missing values.

6. Put variables on roughly the same numeric scale. This is optional, but we recommend it. We explain what we mean and how to do this below.

7. Enter Mata. Do that by typing `mata` at the Stata command prompt. Do not type a colon after the `mata`. This way, when you make a mistake, you will stay in Mata.

8. Use Mata's `st_view()` function (see [M-5] **st_view( )**) to create matrices based on your Stata dataset. Create all the matrices you want or find convenient. The matrices created by `st_view()` are in fact views onto one copy of the data.

9. Perform your matrix calculations.

If you are reading the entries in the order suggested in [M-0] **intro**, see [M-1] **how** next.

1. Start in Stata; load the data

We will use the `auto` dataset and will fit the regression

$$\text{mpg}_j = b_0 + b_1 \text{weight}_j + b_2 \text{foreign}_j + e_j$$

by using the formulas

$$\mathbf{b} = (\mathbf{X}'\mathbf{X})^{-1}\mathbf{X}'\mathbf{y}$$
$$\mathbf{V} = s^2(\mathbf{X}'\mathbf{X})^{-1}$$

where

$$s^2 = \mathbf{e}'\mathbf{e}/(n-k)$$
$$\mathbf{e} = \mathbf{y} - \mathbf{X}\mathbf{b}$$
$$n = \mathrm{rows}(\mathbf{X})$$
$$k = \mathrm{cols}(\mathbf{X})$$

We begin by typing

```
. sysuse auto
(1978 Automobile Data)
```

2. Create any time-series variables

We do not have any time-series variables but, just for a minute, let's pretend we did. If our model contained lagged gnp, we would type

```
. gen lgnp = l.gnp
```

so that we would have a new variable lgnp that we would use in place of l.gnp in the subsequent steps.

3. Create a constant variable

```
. gen cons = 1
```

4. Drop unnecessary variables

We will need the variables mpg, weight, foreign, and cons, so it is easier for us to type keep instead of drop:

```
. keep mpg weight foreign cons
```

5. Drop observations with missing values

We do not have any missing values in our data, but let's pretend we did, or let's pretend we are uncertain. Here is an easy trick for getting rid of observations with missing values:

```
. regress mpg weight foreign cons
```
(output omitted)
```
. keep if e(sample)
```

We estimated a regression using all the variables and then kept the observations regress chose to use. It does not matter which variable you choose as the dependent variable, nor the order of the independent variables, so we just as well could have typed

```
. regress weight mpg foreign cons
```
(output omitted)
```
. keep if e(sample)
```

or even

```
. regress cons mpg weight foreign
(output omitted )
. keep if e(sample)
```

The output produced by `regress` is irrelevant, even if some variables are dropped. We are merely borrowing `regress`'s ability to identify the subsample with no missing values.

Using `regress` causes Stata to make many unnecessary calculations and, if that offends you, here is a more sophisticated alternative:

```
. local 0 "mpg weight foreign cons"
. syntax varlist
. marksample touse
. keep if `touse'
. drop `touse'
```

Using `regress` is easier.

6. Put variables on roughly the same numeric scale

This step is optional, but we recommend it. You are about to use formulas that have been derived by people who assumed that the usual rules of arithmetic hold, such as $(a + b) - c = a + (b - c)$. Many of the standard rules, such as the one shown, are violated when arithmetic is performed in finite precision, and this leads to roundoff error in the final, calculated results.

You can obtain a lot of protection by making sure that your variables are on roughly the same scale, by which we mean their means and standard deviations are all roughly equal. By roughly equal, we mean equal up to a factor of 1,000 or so. So let's look at our data:

```
. summarize
```

Variable	Obs	Mean	Std. Dev.	Min	Max
mpg	74	21.2973	5.785503	12	41
weight	74	3019.459	777.1936	1760	4840
foreign	74	.2972973	.4601885	0	1
cons	74	1	0	1	1

Nothing we see here bothers us much. Variable `weight` is the largest, with a mean and standard deviation that are 1,000 times larger than those of the smallest variable, `foreign`. We would feel comfortable, but only barely, ignoring scale differences. If `weight` were 10 times larger, we would begin to be concerned, and our concern would grow as `weight` grew.

The easiest way to address our concern is to divide `weight` so that, rather than measuring weight in pounds, it measures weight in thousands of pounds:

```
. replace weight = weight / 1000
. summarize
```

Variable	Obs	Mean	Std. Dev.	Min	Max
mpg	74	21.2973	5.785503	12	41
weight	74	3.019459	.7771936	1.76	4.84
foreign	74	.2972973	.4601885	0	1
cons	74	1	0	1	1

What you are supposed to do is make the means and standard deviations of the variables roughly equal. If `weight` had a large mean and reasonable standard deviation, we would have subtracted, so that we would have had a variable measuring weight in excess of some number of pounds. Or we could do both, subtracting, say, 2,000 and then dividing by 100, so we would have weight in excess of 2,000 pounds, measured in 100-pound units.

Remember, the definition of roughly equal allows lots of leeway, so you do not have to give up easy interpretation.

7. Enter Mata

We type

```
. mata
                                                          ─── mata (type end to exit) ───
: _
```

Mata uses a colon prompt, whereas Stata uses a period.

8. Use st_view( ) to access your data

Our matrix formulas are

$$\mathbf{b} = (\mathbf{X'X})^{-1}\mathbf{X'y}$$
$$\mathbf{V} = s^2(\mathbf{X'X})^{-1}$$

where

$$s^2 = \mathbf{e'e}/(n-k)$$
$$\mathbf{e} = \mathbf{y} - \mathbf{Xb}$$
$$n = \text{rows}(\mathbf{X})$$
$$k = \text{cols}(\mathbf{X})$$

so we are going to need **y** and **X**. **y** is an $n \times 1$ column vector of dependent-variable values, and **X** is an $n \times k$ matrix of the k independent variables, including the constant. Rows are observations, columns are variables.

We make the vector and matrix as follows:

```
: st_view(y=., ., "mpg")
: st_view(X=., ., ("weight", "foreign", "cons"))
```

Let us explain. We wish we could type

```
: y = st_view(., "mpg")
: X = st_view(., ("weight", "foreign", "cons"))
```

because that is what the functions are really doing. We cannot because `st_view()` (unlike all other Mata functions), returns a special kind of matrix called a view. A view acts like a regular matrix in nearly every respect, but views do not consume nearly as much memory, because they are in fact views onto the underlying Stata dataset!

We could instead create y and X with Mata's `st_data()` function (see [M-5] **st_data( )**), and then we could type the creation of y and X the natural way,

```
: y = st_data(., "mpg")
: X = st_data(., ("weight", "foreign", "cons"))
```

st_data() returns a real matrix, which is a copy of the data Stata has stored in memory.

We could use st_data() and be done with the problem. For our automobile-data example, that would be a fine solution. But were the automobile data larger, you might run short of memory, and views can save lots of memory. You can create views willy-nilly—lots and lots of them—and never consume much memory! Views are wonderfully convenient and it is worth mastering the little bit of syntax to use them.

st_view() requires three arguments: the name of the view matrix to be created, the observations (rows) the matrix is to contain, and the variables (columns). If we wanted to create a view matrix Z containing all the observations and all the variables, we could type

```
: st_view(Z, ., .)
```

st_view() understands missing value in the second and third positions to mean all the observations and all the variables. Let's try it:

```
: st_view(Z, ., .)
                    err:<istmt>:   3499  Z not found
r(3499);

: _
```

That did not work because Mata requires Z to be predefined. The reasons are technical, but it should not surprise you that function arguments need to be defined before a function can be used. Mata just does not understand that st_view() really does not need Z defined. The way around Mata's confusion is to define Z and then let st_view() redefine it:

```
: Z = .
: st_view(Z, ., .)
```

You can, if you wish, combine all that into one statement

```
: st_view(Z=., ., .)
```

and that is what we did when we defined y and X:

```
: st_view(y=., ., "mpg")
: st_view(X=., ., ("weight", "foreign", "cons"))
```

The second argument (.) specified that we wanted all the observations, and the third argument specified the variables we wanted. Be careful not to omit the "extra" parentheses when typing the variables. Were you to type

```
: st_view(X=., ., "weight", "foreign", "cons")
```

you would be told you typed an invalid expression. st_view() expects three arguments, and the third argument is a row vector specifying the variables to be selected: ("weight", "foreign", "cons").

At this point, we suggest you type

```
: y
(output omitted)
: X
(output omitted)
```

to see that y and X really do contain our data. In case you have lost track of what we have typed, here is our complete session so far:

```
. sysuse auto
. gen cons = 1
. keep mpg weight foreign cons
. regress mpg weight foreign cons
. keep if e(sample)
. replace weight = weight/1000
. mata
: st_view(y=., ., "mpg")
: st_view(X=., ., ("weight", "foreign", "cons"))
```

9. Perform your matrix calculations

To remind you: our matrix calculations are

$$\mathbf{b} = (\mathbf{X}'\mathbf{X})^{-1}\mathbf{X}'\mathbf{y}$$
$$\mathbf{V} = s^2(\mathbf{X}'\mathbf{X})^{-1}$$

where

$$s^2 = \mathbf{e}'\mathbf{e}/(n-k)$$
$$\mathbf{e} = \mathbf{y} - \mathbf{X}\mathbf{b}$$
$$n = \text{rows}(\mathbf{X})$$
$$k = \text{cols}(\mathbf{X})$$

Let's get our regression coefficients,

```
: b = invsym(X'X)*X'y
: b
```

	1
1	-6.587886358
2	-1.650029004
3	41.67970227

and let's form the residuals, define n and k, and obtain s^2,

```
: e  = y - X*b
: n  = rows(X)
: k  = cols(X)
: s2 = (e'e)/(n-k)
```

so we are able to calculate the variance matrix:

```
: V = s2*invsym(X'X)
: V
[symmetric]
```

	1	2	3
1	.4059128628		
2	.4064025078	1.157763273	
3	-1.346459802	-1.57131579	4.689594304

We are done.

We can present the results in more readable fashion by pulling the diagonal of V and calculating the square root of each element:

```
: se = sqrt(diagonal(V))
: (b, se)
                   1               2

    1     -6.587886358      .6371129122
    2     -1.650029004     1.075994086
    3      41.67970227     2.165547114
```

You know that if we were to type

```
: 2+3
  5
```

Mata evaluates the expression and shows us the result, and that is exactly what happened when we typed

```
: (b, se)
```

(b, se) is an expression, and Mata evaluated it and displayed the result. The expression means to form the matrix whose first column is b and second column is se. We could obtain a listing of the coefficient, standard error, and its t statistic by asking Mata to display (b, se, b:/se),

```
: (b, se, b:/se)
                   1               2                3

    1     -6.587886358      .6371129122     -10.34021793
    2     -1.650029004     1.075994086      -1.533492633
    3      41.67970227     2.165547114       19.24673077
```

In the expression above, b:/se means to divide the elements of b by the elements of se. :/ is called a colon operator and you can learn more about it by seeing [M-2] **op_colon**.

We could add the significance level by typing

```
: (b, se, b:/se, 2*ttail(n-k, abs(b:/se)))
                   1               2                3               4

    1     -6.587886358      .6371129122     -10.34021793     8.28286e-16
    2     -1.650029004     1.075994086      -1.533492633     .1295987129
    3      41.67970227     2.165547114       19.24673077     6.89556e-30
```

Those are the same results reported by regress; type

```
. sysuse auto
. replace weight = weight/1000
. regress mpg weight foreign
```

and compare results.

Review

Our complete session was

```
. sysuse auto
. gen cons = 1
. keep mpg weight foreign cons
. regress mpg weight foreign cons
. keep if e(sample)
. replace weight = weight/1000

. mata
: st_view(y=., ., "mpg")
: st_view(X=., ., ("weight", "foreign", "cons"))

: b = invsym(X'X)*X'y
: b
: e = y - X*b
: n = rows(X)
: k = cols(X)
: s2= (e'e)/(n-k)
: V = s2*invsym(X'X)
: V

: se = sqrt(diagonal(V))
: (b, se)
: (b, se, b:/se)
: (b, se, b:/se, 2*ttail(n-k, abs(b:/se)))
: end
```

Reference

Gould, W. 2006. Mata Matters: Interactive use. *Stata Journal* 6: 387–396.

Also See

[M-1] **intro** — Introduction and advice

Title

[M-1] **LAPACK** — The LAPACK linear-algebra routines

Description

LAPACK stands for Linear Algebra PACKage and is a freely available set of FORTRAN 77 routines for solving systems of simultaneous equations, eigenvalue problems, and singular value problems. Many of the LAPACK routines are based on older EISPACK and LINPACK routines, and the more modern LAPACK does much of its computation by using BLAS (Basic Linear Algebra Subprogram).

Remarks

The LAPACK and BLAS routines form the basis for many of Mata's linear-algebra capabilities. Individual functions of Mata that use LAPACK routines always make note of that fact.

For up-to-date information on LAPACK, see http://www.netlib.org/lapack/.

Acknowledgments

We thank the authors of LAPACK for their excellent work:

Anderson, E., Z. Bai, C. Bischof, S. Blackford, J. Demmel, J. Dongarra, J. Du Croz, A. Greenbaum, S. Hammarling, A. McKenney, and D. Sorensen. 1999. *LAPACK Users' Guide*, 3rd ed. Philadelphia, PA: Society for Industrial and Applied Mathematics.

Also See

[M-1] **intro** — Introduction and advice

Title

Summary

Limits:

	minimum	maximum
Scalars, vectors, matrices		
rows	0	2,147,483,647
columns	0	2,147,483,647
String elements, length	0	2,147,483,647

Stata's `matsize` plays no role in these limits.

Size approximations:

	memory requirements
real matrices	$oh + r*c*8$
complex matrices	$oh + r*c*16$
pointer matrices	$oh + r*c*8$
string matrices	$oh + r*c*8 + total\_length\_of\_strings$

where r and c represent the number of rows and columns and
where oh is overhead and is approximately 64 bytes

Description

Mata imposes limits, but those limits are of little importance compared with the memory requirements. Mata stores matrices in memory and requests the memory for them from the operating system.

Remarks

Mata requests (and returns) memory from the operating system as it needs it, and if the operating system cannot provide it, Mata issues the following error:

```
: x = foo(A, B)
            foo():  3900   unable to allocate ...
            <istmt>:    -  function returned error
    r(3900);
```

Stata's `matsize` (see [R] **matsize**) and Stata's `set memory` value (see [D] **memory**) play no role in Mata or, at least, they play no direct role. `set memory` specifies the amount of memory to be set aside for the storage of Stata's dataset; hence, larger values of `set memory` leave less memory for Mata.

Also See

Title

[M-1] naming — Advice on naming functions and variables

Syntax

A *name* is 1–32 characters long, the first character of which must be

> A–Z   a–z   _

and the remaining characters of which may be

> A–Z   a–z   _   0–9

except that names may not be a word reserved by Mata (see [M-2] **reswords** for a list).

Examples of names include

> x   x2   alpha
> logarithm_of_x   LogOfX

Case matters: `alpha`, `Alpha`, and `ALPHA` are different names.

Variables and functions have separate name spaces, which means a variable and a function can have the same name, such as `value` and `value()`, and Mata will not confuse them.

Description

Advice is offered on how to name variables and functions.

Remarks

Remarks are presented under the following headings:

> *Interactive use*
> *Naming variables*
> *Naming functions*
> *What happens when functions have the same names*
> *How to determine if a function name has been taken*

Interactive use

Use whatever names you find convenient: Mata will tell you if there is a problem.

The following sections are for programmers who want to write code that will require the minimum amount of maintenance in the future.

49

Naming variables

Mostly, you can name variables however you please. Variables are local to the program in which they appear, so one function can have a variable or argument named x and another function can have a variable or argument of the same name, and there is no confusion.

If you are writing a large system of programs that has global variables, on the other hand, we recommend that you give the global variables long names, preferably with a common prefix identifying your system. For instance,

```
multeq_no_of_equations
multeq_eq
multeq_inuse
```

This way, your variables will not become confused with variables from other systems.

Naming functions

Our first recommendation is that, for the most part, you give functions all-lowercase names: foo() rather than Foo() or FOO(). If you have a function with one or more capital letters in the name, and if you want to save the function's object code, you must do so in .mlib libraries; .mo files are not sufficient. .mo files require that filename be the same as the function name, which means Foo.mo for Foo(). Not all operating systems respect case in filenames. Even if your operating system does respect case, you will not be able to share your .mo files with others whose operating systems do not.

We have no strong recommendation against mixed case; we merely remind you to use .mlib library files if you use it.

Of greater importance is the name you choose. Mata provides many functions and more will be added over time. You will find it more convenient if you choose names that StataCorp and other users do not choose.

That means to avoid words that appear in the English-language dictionary and to avoid short names, say, those four characters or fewer. You might have guessed that svd() would be taken, but who would have guessed lud()? Or qrd()? Or e()?

Your best defense against new official functions, and other user-written functions, is to choose long function names.

What happens when functions have the same names

There are two kinds of official functions: built-in functions and library functions. User-written functions are invariably library functions (here we draw no distinction between functions supplied in .mo files and those supplied in .mlib files).

Mata will issue an error message if you attempt to define a function with the same name as a built-in function.

Mata will let you define a new function with the same name as a library function if the library function is not currently in memory. If you store your function in a .mo file or a .mlib library, however, in the future the official Mata library function will take precedence over your function: your function will never be loaded. This feature works nicely for interactive users, but for long-term programming, you will want to avoid naming your functions after Mata functions.

A similar result is obtained if you name your function after a user-written function that is installed on your computer. You can do so if the user-written function is not currently in memory. In the future, however, one or the other function will take precedence and, no matter which, something will break.

How to determine if a function name has been taken

Use mata which (see [M-3] **mata which**):

```
: mata which det_of_triangular()
function det_of_triangular() not found
r(111);
: mata which det()
  det():  lmatabase.mlib
```

Also See

[M-2] **reswords** — Reserved words

[M-1] **intro** — Introduction and advice

Title

Syntax

Action	Permutation matrix notation	Permutation vector notation
permute rows	$B = P*A$	$B = A[p,.]$
permute columns	$B = A*P$	$B = A[.,p]$
unpermute rows	$B = P'A$	$B = A \; ; B[p,.] = A$
		or
		$B = A[\mathtt{invorder}(p),.]$
unpermute columns	$B = A*P'$	$B = A \; ; B[.,p] = A$
		or
		$B = A[., \mathtt{invorder}(p)]$

A *permutation matrix* is an $n \times n$ matrix that is a row (or column) permutation of the identity matrix.

A *permutation vector* is a $1 \times n$ or $n \times 1$ vector of the integers 1 through n.

The following permutation matrix and permutation vector are equivalent:

$$P = \begin{bmatrix} 0 & 1 & 0 \\ 0 & 0 & 1 \\ 1 & 0 & 0 \end{bmatrix} \iff p = \begin{bmatrix} 2 \\ 3 \\ 1 \end{bmatrix}$$

Either can be used to permute the rows of

$$A = \begin{bmatrix} a & b & c & d \\ e & f & g & h \\ i & j & k & l \end{bmatrix} \quad \text{to produce} \quad \begin{bmatrix} e & f & g & h \\ i & j & k & l \\ a & b & c & d \end{bmatrix}$$

and to permute the columns of

$$P = \begin{bmatrix} m & n & o \\ p & q & r \\ s & t & u \\ v & w & x \end{bmatrix} \quad \text{to produce} \quad \begin{bmatrix} n & o & m \\ q & r & p \\ t & u & s \\ w & x & v \end{bmatrix}$$

Permutation matrices are a special kind of orthogonal matrix that, via multiplication, reorder the rows or columns of another matrix. Permutation matrices cast the reordering in terms of multiplication.

Permutation vectors also reorder the rows or columns of another matrix, but they do it via subscripting. This alternative method of achieving the same end is, computerwise, more efficient, in that it uses less memory and less computer time.

The relationship between the two is shown below.

Remarks

Remarks are presented under the following headings:

> *Permutation matrices*
> *How permutation matrices arise*
> *Permutation vectors*

Permutation matrices

A permutation matrix is a square matrix whose rows are a permutation of the identity matrix. The following are the full set of all 2×2 permutation matrices:

$$\begin{bmatrix} 1 & 0 \\ 0 & 1 \end{bmatrix} \tag{1}$$

$$\begin{bmatrix} 0 & 1 \\ 1 & 0 \end{bmatrix} \tag{2}$$

Let P be an $n \times n$ permutation matrix. If $n \times m$ matrix A is premultiplied by P, the result is to reorder its rows. For example,

$$\begin{array}{ccccc} P & * & A & = & PA \\ \begin{bmatrix} 0 & 1 & 0 \\ 0 & 0 & 1 \\ 1 & 0 & 0 \end{bmatrix} & * & \begin{bmatrix} 1 & 2 & 3 \\ 4 & 5 & 6 \\ 7 & 8 & 9 \end{bmatrix} & = & \begin{bmatrix} 4 & 5 & 6 \\ 7 & 8 & 9 \\ 1 & 2 & 3 \end{bmatrix} \end{array} \tag{3}$$

Above, we illustrated the reordering using square matrix A, but A did not have to be square.

If $m \times n$ matrix B is postmultiplied by P, the result is to reorder its columns. We illustrate using square matrix A again:

$$\begin{array}{ccccc} A & * & P & = & AP \\ \begin{bmatrix} 1 & 2 & 3 \\ 4 & 5 & 6 \\ 7 & 8 & 9 \end{bmatrix} & * & \begin{bmatrix} 0 & 1 & 0 \\ 0 & 0 & 1 \\ 1 & 0 & 0 \end{bmatrix} & = & \begin{bmatrix} 3 & 1 & 2 \\ 6 & 4 & 5 \\ 9 & 7 & 8 \end{bmatrix} \end{array} \tag{4}$$

Say that we reorder the rows of A by forming PA. Obviously, we can unreorder the rows by forming $P^{-1}PA$. Because permutation matrices are orthogonal, their inverses are equal to their transpose. Thus the inverse of the permutation matrix $(0,1,0\backslash 0,0,1\backslash 1,0,0)$ we have been using is $(0,0,1\backslash 1,0,0\backslash 0,1,0)$. For instance, taking our results from (3)

$$P' \quad * \quad PA \quad = \quad A$$

$$\begin{bmatrix} 0 & 0 & 1 \\ 1 & 0 & 0 \\ 0 & 1 & 0 \end{bmatrix} * \begin{bmatrix} 4 & 5 & 6 \\ 7 & 8 & 9 \\ 1 & 2 & 3 \end{bmatrix} = \begin{bmatrix} 1 & 2 & 3 \\ 4 & 5 & 6 \\ 7 & 8 & 9 \end{bmatrix} \tag{3'}$$

Allow us to summarize:

1. A permutation matrix P is a square matrix whose rows are a permutation of the identity matrix.

2. $PA =$ a row permutation of A.

3. $AP =$ a column permutation of A.

4. The inverse permutation is given by P'.

5. $P'PA = A$.

6. $APP' = A$.

How permutation matrices arise

Some of Mata's matrix functions implicitly permute the rows (or columns) of a matrix. For instance, the LU decomposition of matrix A is defined as

$$A = LU$$

where L is lower triangular and U is upper triangular. For any matrix A, one can solve for L and U, and Mata has a function that will do that (see [M-5] **lud( )**). However, Mata's function does not solve the problem as stated. Instead, it solves
$$P'A = LU$$

where P' is a permutation matrix that Mata makes up! Just to be clear; Mata's function solves for L and U, but for a row permutation of A, not A itself, although the function does tell you what permutation it chose (the function returns L, U, and P). The function permutes the rows because, that way, it can produce a more accurate answer.

You will sometimes read that a function engages in pivoting. What that means is that, rather than solving the problem for the matrix as given, it solves the problem for a permutation of the original matrix, and the function chooses the permutation in a way to minimize numerical roundoff error. Functions that do this invariably return the permutation matrix along with the other results, because you are going to need it.

For instance, one use of LU decomposition is to calculate inverses. If $A = LU$ then $A^{-1} = U^{-1}L^{-1}$. Calculating the inverses of triangular matrices is an easy problem, so one recipe for calculating inverses is

1. decompose A into L and U,

2. calculate U^{-1},

3. calculate L^{-1}, and

4. multiply the results together.

That would be the solution except that the LU decomposition function does not decompose A into L and U; it decomposes $P'A$, although the function does tell us P. Thus we can write,

$$P'A = LU$$
$$A = PLU \qquad \text{(remember } P'^{-1} = P\text{)}$$
$$A^{-1} = U^{-1}L^{-1}P'$$

Thus the solution to our problem is to use the U and L just as we planned—calculate $U^{-1}L^{-1}$—and then make a column permutation of that, which we can do by postmultiplying by P'.

There is, however, a detail that we have yet to reveal to you: Mata's LU decomposition function does not return P, the permutation matrix. It instead returns p, a permutation vector equivalent to P, and so the last step—forming the column permutation by postmultiplying by P'—is done differently. That is the subject of the next section.

Permutation vectors are more efficient than permutation matrices, but you are going to discover that they are not as mathematically transparent. Thus when working with a function that returns a permutation vector—when working with a function that permutes the rows or columns—think in terms of permutation matrices and then translate back into permutation vectors.

Permutation vectors

Permutation vectors are used with Mata's subscripting operator, so before explaining permutation vectors, let's understand subscripting.

Not surprisingly, Mata allows subscripting. Given matrix

$$A = \begin{bmatrix} 1 & 2 & 3 \\ 4 & 5 & 6 \\ 7 & 8 & 9 \end{bmatrix}$$

Mata understands that

$$A[2,3] = 6$$

Mata also understands that if one or the other subscript is specified as . (missing value), the entire column or row is to be selected:

$$A[.,3] = \begin{bmatrix} 3 \\ 6 \\ 9 \end{bmatrix}$$

$$A[2,.] = \begin{bmatrix} 4 & 5 & 6 \end{bmatrix}$$

Mata also understands that if a vector is specified for either subscript

$$A\begin{bmatrix} 2 \\ 3 \\ 2 \end{bmatrix}, .] = \begin{bmatrix} 4 & 5 & 6 \\ 7 & 8 & 9 \\ 4 & 5 & 6 \end{bmatrix}$$

In Mata, we would actually write the above as $A[(2\backslash3\backslash2),.]$, and Mata would understand that we want the matrix made up of rows 2, 3, and 2 of A, all columns. Similarly, we can request all rows, columns 2, 3, and 2:

$$A[.,(2,3,2)] = \begin{bmatrix} 2 & 3 & 2 \\ 5 & 6 & 5 \\ 8 & 9 & 8 \end{bmatrix}$$

In the above, we wrote (2,3,2) as a row vector because it seems more logical that way, but we could just as well have written $A[.,(2\backslash3\backslash2)]$. In subscripting, Mata does not care whether the vectors are rows or columns.

In any case, we can use a vector of indices inside Mata's subscripts to select rows and columns of a matrix, and that means we can permute them. All that is required is that the vector we specify contain a permutation of the integers 1 through n because, otherwise, we would repeat some rows or columns and omit others.

A permutation vector is an $n \times 1$ or $1 \times n$ vector containing a permutation of the integers 1 through n. For example, the permutation vector equivalent to the permutation matrix

$$P = \begin{bmatrix} 0 & 1 & 0 \\ 0 & 0 & 1 \\ 1 & 0 & 0 \end{bmatrix}$$

is

$$p = \begin{bmatrix} 2 \\ 3 \\ 1 \end{bmatrix}$$

p can be used with subscripting to permute the rows of A

$$A[p,.] = \begin{bmatrix} 4 & 5 & 6 \\ 7 & 8 & 9 \\ 1 & 2 & 3 \end{bmatrix}$$

and similarly, $A[.,p]$ would permute the columns.

Also subscripting can be used on the left-hand side of the equal-sign assignment operator. So far, we have assumed that the subscripts are on the right-hand side of the assignment operator, such as

```
B = A[p,.]
```

We have learned that if $p = (2\backslash3\backslash1)$ (or $p = (2,3,1)$), the result is to copy the second row of A to the first row of B, the third row of A to the second row of B, and the first row of A to the third row of B. Coding

```
B[p,.] = A
```

does the inverse: it copies the first row of A to the second row of B, the second row of A to the third row of B, and the third row of A to the first row of B. $B[p,.]=A$ really is the inverse of $C=A[p,.]$ in that, if we code

```
C = A[p,.]
B[p,.] = C
```

B will be equal to *C*, and if we code

```
C[p,.] = A
B = C[p,.]
```

B will also be equal to *C*.

There is, however, one pitfall that you must watch for when using subscripts on the left-hand side: the matrix on the left-hand side must already exist and it must be of the appropriate (here same) dimension. Thus when performing the inverse copy, it is common to code

```
B = C
B[p,.] = C
```

The first line is not unnecessary; it is what ensures that *B* exists and is of the proper dimension, although we could just as well code

```
B = J(rows(C), cols(C), .)
B[p,.] = C
```

The first construction is preferred because it ensures that *B* is of the same type as *C*. If you really like the second form, you should code

```
B = J(rows(C), cols(C), missingof(C))
B[p,.] = C
```

Going back to the preferred code

```
B = C
B[p,.] = C
```

some programmers combine it into one statement:

```
(B=C)[p,.] = C
```

Also Mata provides an `invorder`(*p*) (see [M-5] **invorder( )**) that will return an inverted *p* appropriate for use on the right-hand side, so you can also code

```
B = C[invorder(p),.]
```

Also See

Title

[M-1] returnedargs — Function arguments used to return results

Syntax

$y = f(x, \ldots)$ (function returns result the usual way)

$g(x, \ldots, y)$ (function returns result in argument y)

Description

Most Mata functions leave their arguments unchanged and return a result:

    : $y = f(x, \ldots)$

Some Mata functions, however, return nothing and instead return results in one or more arguments:

    : $g(x, \ldots, y)$

If you use such functions interactively and the arguments that are to receive results are not already defined (y in the above example), you will get a variable-not-found error. The solution is to define the arguments to contain something—anything—before calling the function:

    : $y = .$
    : $g(x, \ldots, y)$

You can combine this into one statement:

    : $g(x, \ldots, y=.)$

Remarks

$sqrt(a)$—see [M-5] **sqrt( )**—calculates the (element-by-element) square root of a and returns the result:

```
: x = 4
: y = sqrt(x)
: y              // y now contains 2
  2
: x              // x is unchanged
  4
```

Most functions work like `sqrt()`, although many take more than one argument.

On the other hand, $polydiv(c_a, c_b, c_q, c_r)$—see [M-5] **polyeval( )**—takes the polynomial stored in c_a and the polynomial stored in c_b and divides them. It returns the quotient in the third argument (c_q) and the remainder in the fourth (c_r). c_a and c_b are left unchanged. The function itself returns nothing:

```
: A = (1,2,3)
: B = (0,1)
: polydiv(A, B, Q, R)
```

```
: Q                          // Q has been redefined
          1    2

    1   [ 2    3 ]

: R                          // as has R
    1
: A                          // while A and B are unchanged
          1    2    3

    1   [ 1    2    3 ]

: B
          1    2

    1   [ 0    1 ]
```

As another example, st_view(V, i, j)—see [M-5] **st_view( )**—creates a view onto the Stata dataset. Views are like matrices but consume less memory. Arguments i and j specify the observations and variables to be selected. Rather than returning the matrix, however, the result is returned in the first argument (V).

```
: st_view(V, (1\5), ("mpg", "weight"))
: V
          1        2

    1   [ 22      2930 ]
    2   [ 15      4080 ]
```

If you try to use these functions interactively, you will probably get an error:

```
: polydiv(A, B, Q, R)
              <istmt>:   3499   Q not found
r(3499);
: st_view(V, (1\5), ("mpg", "weight"))
              <istmt>:   3499   V not found
r(3499);
```

Arguments must be defined before they are used, even if their only purpose is to receive a newly calculated result. In such cases, it does not matter how the argument is defined because its contents will be replaced. Easiest is to fill in a missing value:

```
: Q = .
: R = .
: polydiv(A, B, Q, R)
: V = .
: st_view(V, (1\5), ("mpg", "weight"))
```

You can also define the argument inside the function:

```
: polydiv(A, B, Q=., R=.)
: st_view(V=., (1\5), ("mpg", "weight"))
```

When you use functions like these inside a program, however, you need not worry about defining the arguments, because they are defined by virtue of appearing in your program:

```
function foo()
{
    ...
    polydiv(A, B, Q, R)
    st_view(V, (1\5), ("mpg", "weight"))
    ...
}
```

When Mata compiles your program, however, you may see warning messages:

```
: function foo()
> {
>     ...
>     polydiv(A, B, Q, R)
>     st_view(V, (1\5), ("mpg", "weight"))
>     ...
> }
note: variable Q may be used before set
note: variable R may be used before set
note: variable V may be used before set
```

If the warning messages bother you, either define the variables before they are used just as you would interactively or use pragma to suppress the warning messages; see [M-2] **pragma**.

Also See

[M-1] **intro** — Introduction and advice

Title

> **[M-1] source** — Viewing the source code

Syntax

> . viewsource *functionname*.mata

Description

Many Mata functions are written in Mata. viewsource will allow you to examine their source code.

Remarks

Some Mata functions are implemented in C (they are part of Mata itself), and others are written in Mata.

viewsource allows you to look at the official source code written in Mata. Reviewing this code is a great way to learn Mata.

The official source code is stored in .mata files. To see the source code for diag(), for instance, type

> . viewsource diag.mata

You type this at Stata's dot prompt, not at Mata's colon prompt.

If a function is built in, such as abs(), here is what will happen when you attempt to view the source code:

```
. viewsource abs.mata
file "abs.mata" not found
r(601);
```

You can verify that abs() is built in by using the mata which (see [M-3] **mata which**) command:

```
. mata: mata which abs()
  abs():  built-in
```

viewsource can be also used to look at source code of user-written functions if the distribution included the source code (it might not).

Also See

[P] **viewsource** — View source code

[M-1] **intro** — Introduction and advice

Title

Syntax

$somefunction(\ldots,\ real\ scalar\ tol,\ \ldots)$

where, concerning argument *tol*,

optional   Argument *tol* is usually optional; not specifying *tol* is equivalent to specifying $tol = 1$.

tol > 0   Specifying $tol > 0$ specifies the amount by which the usual tolerance is to be multiplied: $tol = 2$ means twice the usual tolerance; $tol = .5$ means half the usual tolerance.

tol < 0   Specifying $tol < 0$ specifies the negative of the value to be used for the tolerance: $tol = -1e-14$ means $1e-14$ is to be used.

tol = 0   Specifying $tol = 0$ means all numbers are to be taken at face value, no matter how close to 0 they are. The single exception is when *tol* is applied to values that, mathematically, must be greater than or equal to zero. Then negative values (which arise from roundoff error) are treated as if they were zero.

The default tolerance is given by formula, such as

$eta = \texttt{1e-14}$

or

$eta = \texttt{epsilon(1)}$ (see [M-5] **epsilon( )**)

or

$eta = \texttt{1000*epsilon(trace(abs(}A\texttt{))/rows(}A\texttt{))}$

Specifying $tol > 0$ specifies a value to be used to multiply *eta*. Specifying $tol < 0$ specifies that $-tol$ be used in place of *eta*. Specifying $tol = 0$ specifies that *eta* be set to 0.

The formula for *eta* and how *eta* is used are found under *Remarks*. For instance, the *Remarks* might say that A is declared to be singular if any diagonal element of U of its LU decomposition is less than or equal to *eta*.

Description

The results provided by many of the numerical routines in Mata depend on tolerances. Mata provides default tolerances, but those can be overridden.

Remarks

Remarks are presented under the following headings:

The problem
Absolute versus relative tolerances
Specifying tolerances

The problem

In many formulas, zero is a special number in that, when the number arises, sometimes the result cannot be calculated or, other times, something special needs to be done.

The problem is that zero—0.00000000000—seldom arises in numerical calculation. Because of roundoff error, what would be zero were the calculation performed in infinite precision in fact is $1.03948e-15$, or $-4.4376e-16$, etc.

If one behaves as if these small numbers are exactly what they seem to be ($1.03948e-15$ is taken to mean $1.03948e-15$ and not zero), some formulas produce wildly inaccurate results; see [M-5] **lusolve( )** for an example.

Thus routines use *tolerances*—preset numbers—to determine when a number is small enough to be considered to be zero.

The problem with tolerances is determining what they ought to be.

Absolute versus relative tolerances

Tolerances come in two varieties: absolute and relative.

An absolute tolerance is a fixed number that is used to make direct comparisons. If the tolerance for a particular routine were $1e-14$, then $8.99e-15$ in some calculation would be considered to be close enough to zero to act as if it were, in fact, zero, and $1.000001e-14$ would be considered a valid, nonzero number.

But is $1e-14$ small? The number may look small to you, but whether $1e-14$ is small depends on what is being measured and the units in which it is measured. If all the numbers in a certain problem were around $1e-12$, you might suspect that $1e-14$ is a reasonable number.

That leads to relative measures of tolerance. Rather than treating, say, a predetermined quantity as being so small as to be zero, one specifies a value (e.g., $1e-14$) multiplied by something and uses that as the definition of small.

Consider the following matrix:

$$\begin{bmatrix} 5.5e-15 & 1.2e-16 \\ 1.3e-16 & 6.4e-15 \end{bmatrix}$$

What is the rank of the matrix? One way to answer that question would be to take the LU decomposition of the matrix and then count the number of diagonal elements of U that are greater than zero. Here, however, we will just look at the matrix.

The absolutist view is that the matrix is full of roundoff error and that the matrix is really indistinguishable from the matrix

$$\begin{bmatrix} 0 & 0 \\ 0 & 0 \end{bmatrix}$$

The matrix has rank 0. The relativist view is that the matrix has rank 2 because, other than a scale factor of $1e-16$, the matrix is indistinguishable from

$$\begin{bmatrix} 55.0 & 1.2 \\ 1.3 & 64.0 \end{bmatrix}$$

There is no way this question can be answered until someone tells you how the matrix arose and the units in which it is measured.

Nevertheless, most Mata routines would (by default) adopt the relativist view: the matrix is of full rank. That is because most Mata routines are implemented using relative measures of tolerance, chosen because Mata routines are mostly used by people performing statistics, who tend to make calculations such as $X'X$ and $X'Z$ on data matrices, and those resulting matrices can contain very large numbers. Such a matrix might contain

$$\begin{bmatrix} 5.5e+14 & 1.2e+12 \\ 1.3e+13 & 2.4e+13 \end{bmatrix}$$

Given a matrix with such large elements, one is tempted to change one's view as to what is small. Calculate the rank of the following matrix:

$$\begin{bmatrix} 5.5e+14 & 1.2e+12 & 1.5e-04 \\ 1.3e+13 & 2.4e+13 & 2.8e-05 \\ 1.3e-04 & 2.4e-05 & 8.7e-05 \end{bmatrix}$$

This time, we will do the problem correctly: we will take the LU decomposition and count the number of nonzero entries along the diagonal of U. For the above matrix, the diagonal of U turns out to be (5.5e+14, 2.4e+13, .000087).

An absolutist would tell you that the matrix is of full rank; the smallest number along the diagonal of U is .000087 (8.7e−5), and that is still a respectable number, at least when compared with computer precision, which is about 2.22e−16 (see [M-5] **epsilon( )**).

Most Mata routines would tell you that the matrix has rank 2. Numbers such as .000087 may seem respectable when compared with machine precision, but .000087 is, relatively speaking, a very small number, being about 4.6e−19 relative to the average value of the diagonal elements.

Specifying tolerances

Most Mata routines use relative tolerances, but there is no rule. You must read the documentation for the function you are using.

When the tolerance entry for a function directs you here, [M-1] **tolerance**, then the tolerance works as summarized under *Syntax* above. Specify a positive number, and that number multiplies the default; specify a negative number, and the corresponding positive number is used in place of the default.

Also See

[M-5] **epsilon( )** — Unit roundoff error (machine precision)

[M-5] **solve_tol( )** — Tolerance used by solvers and inverters

[M-1] **intro** — Introduction and advice

[M-2] Language definition

Title

[M-2] intro — Language definition

Contents

┌─────────────────┐
│ Flow of control │
└─────────────────┘

[M-2] **if**	if (*exp*) ... else ...
[M-2] **for**	for (*exp1*; *exp2*; *exp3*) *stmt*
[M-2] **while**	while (*exp*) *stmt*
[M-2] **do**	do ... while (*exp*)
[M-2] **break**	Break out of for, while, or do loop
[M-2] **continue**	Continue with next iteration of for, while, or do loop
[M-2] **goto**	goto *label*
[M-2] **return**	return and return(*exp*)

┌────────────────┐
│ Special topics │
└────────────────┘

[M-2] **semicolons**	Use of semicolons
[M-2] **void**	Void matrices
[M-2] **pointers**	Pointers
[M-2] **ftof**	Passing functions to functions

┌─────────────┐
│ Error codes │
└─────────────┘

| [M-2] **errors** | Error codes |

Description

This section defines the Mata programming language.

Remarks

[M-2] **syntax** provides an overview, dense and brief, and the other sections expand on it.

Also see [M-1] **intro** for an introduction to Mata.

Also See

[M-0] **intro** — Introduction to the Mata manual

Title

Syntax

```
for, while, or do {
    ...
    if (...) {
        ...
        break
    }
}
stmt                    ← break jumps here
...
```

Description

break exits the innermost for, while, or do loop. Execution continues with the statement immediately following the close of the loop, just as if the loop had terminated normally.

break nearly always occurs following an if.

Remarks

In the following code,

```
for (i=1; i<=rows(A); i++) {
    for (j=1; j<=cols(A); j++) {
        ...
        if (A[i,j]==0) break
    }
    printf("j = %g\n", j)
}
```

the break statement will be executed if any element of A[i,j] is zero. Assume that the statement is executed for i=2 and j=3. Execution will continue with the printf() statement, which is to say, the j loop will be canceled but the i loop will continue. The value of *j* upon exiting the loop will be 3; when you break out of the loop, the j++ is not executed.

Also See

[M-2] **do** — do ... while (exp)

[M-2] **for** — for (exp1; exp2; exp3) stmt

[M-2] **while** — while (exp) stmt

[M-2] **continue** — Continue with next iteration of for, while, or do loop

[M-2] **intro** — Language definition

Title

Syntax

>/* *enclosed comment* */
>
>// *rest-of-line comment*

Notes:

1. Comments may appear in do-files and ado-files; they are not allowed interactively.

2. Stata's beginning-of-the-line asterisk comment is not allowed in Mata:

   >. * *valid in Stata but not in Mata*

Description

/* and */ and // are how you place comments in Mata programs.

Remarks

There are two comment styles: /* and */ and //. You may use one, the other, or both.

Remarks are presented under the following headings:

>The /* */ enclosed comment
>The // rest-of-line comment

The /* */ enclosed comment

Enclosed comments may appear on a line:

```
/* What follows uses an approximation formula: */
```

Enclosed comments may appear within a line and even in the middle of a Mata expression:

```
x = x + /*left-single quote*/ char(96)
```

Enclosed comments may themselves contain multiple lines:

```
/*
    We use the approximation based on sin(x) approximately
    equaling x for small x; x measure in radians
*/
```

Enclosed comments may be nested, which is useful for commenting out code that itself contains comments:

```
/*
for (i=1; i<=rows(x); i++) {          /* normalization */
        x[i] = x[i] :/ value[i]
}
*/
```

The // rest-of-line comment

The rest-of-line comment may appear by itself on a line

    // What follows uses an approximation formula:

or it may appear at the end of a line:

    x = x + char(96)   *//* append single quote

In either case, the comment concludes when the line ends.

Also See

[M-2] **intro** — Language definition

Title

[M-2] continue — Continue with next iteration of for, while, or do loop

Syntax

```
for, while, or do {
    ...
    if (...) {
        ...
        continue
    }
    ...
}
...
```

Description

continue restarts the innermost for, while, or do loop. Execution continues just as if the loop had reached its logical end.

continue nearly always occurs following an if.

Remarks

The following two code fragments are equivalent:

```
for (i=1; i<=rows(A); i++) {
    for (j=1; j<=cols(A); j++) {
        if (i==j) continue
        ... action to be performed on A[i,j] ...
    }
}
```

and

```
for (i=1; i<=rows(A); i++) {
    for (j=1; j<=cols(A); j++) {
        if (i!=j) {
            ... action to be performed on A[i,j] ...
        }
    }
}
```

continue operates on the innermost for or while loop, and even when the continue action is taken, standard end-of-loop processing takes place (which is j++ here).

Also See

[M-2] **do** — do . . . while (exp)

[M-2] **for** — for (exp1; exp2; exp3) stmt

[M-2] **while** — while (exp) stmt

[M-2] **break** — Break out of for, while, or do loop

[M-2] **intro** — Language definition

Title

Syntax

> *declaration*$_1$ *fcnname* (*declaration*$_2$)
> {
>   *declaration*$_3$
>   . . .
> }

such as

```
real matrix myfunction(real matrix X, real scalar i)
{
    real scalar    j, k
    real vector    v

    . . .
}
```

declaration$_1$ is one of

> function
> *type* [function]
> void [function]

declaration$_2$ is

> [*type*] *argname* [, [*type*] *argname* [, ...]]

where *argname* is the name you wish to assign to the argument.

declaration$_3$ are lines of the form of either of

> *type*   *varname* [, *varname* [, ...]]
> external [*type*] *varname* [, *varname* [, ...]]

type is defined as one of

> *eltype orgtype*   such as `real vector`
> *eltype*   such as `real`
> *orgtype*   such as `vector`

74

eltype and *orgtype* are each one of

eltype	*orgtype*
transmorphic	matrix
numeric	vector
real	rowvector
complex	colvector
string	scalar
pointer	

If *eltype* is not specified, `transmorphic` is assumed. If *orgtype* is not specified, `matrix` is assumed.

Description

Types and the use of declarations is explained. Also discussed is the calling convention (functions are called by address, not by value, and so may change the caller's arguments), and the use of external globals.

Mata also has structures—the *eltype* is `struct` *name*—but these are not discussed here. For a discussion of structures, see [M-2] **struct**.

Declarations are optional but, for careful work, their use is recommended.

Remarks

Remarks are presented under the following headings:

> *The purpose of declarations*
> *Types, element types, and organizational types*
> *Implicit declarations*
> *Element types*
> *Organizational types*
> *Function declarations*
> *Argument declarations*
> *The by-address calling convention*
> *Variable declarations*
> *Linking to external globals*

The purpose of declarations

Declarations occur in three places: in front of function definitions, inside the parentheses defining the function's arguments, and at the top of the body of the function, defining private variables the function will use. For instance, consider the function

```
real matrix swaprows(real matrix A, real scalar i1, real scalar i2)
{
    real matrix     B
    real rowvector  v

    B = A
    v = B[i1, .]
    B[i1, .] = B[i2, .]
    B[i2, .] = v
    return(B)
}
```

This function returns a copy of matrix A with rows i1 and i2 swapped.

There are three sets of declarations in the above function. First, there is a declaration in front of the function name:

```
real matrix swaprows(...)
{
        ...
}
```

That declaration states that this function will return a real matrix.

The second set of declarations occur inside the parentheses:

```
... swaprows(real matrix A, real scalar i1, real scalar i2)
{
        ...
}
```

Those declarations state that this function expects to receive three arguments, which we chose to call A, i1, and i2, and which we expect to be a real matrix, a real scalar, and a real scalar, respectively.

The third set of declarations occur at the top of the body of the function:

```
... swaprows(...)
{
        real matrix      B
        real rowvector   v

        ...
}
```

Those declarations state that we will use variables B and v inside our function and that, as a matter of fact, B will be a real matrix and v a real row vector.

We could have omitted all those declarations. Our function could have read

```
function swaprows(A, i1, i2)
{
        B = A
        v = B[i1, .]
        B[i1, .] = B[i2, .]
        B[i2, .] = v
        return(B)
}
```

and it would have worked just fine. So why include the declarations?

1. By including the outside declaration, we announced to other programs what to expect. They can depend on swaprows() returning a real matrix because, when swaprows() is done, Mata will verify that the function really is returning a real matrix and, if it is not, abort execution.

    Without the outside declaration, anything goes. Our function could return a real scalar in one case, a complex row vector in another, and nothing at all in yet another case.

    Including the outside declaration makes debugging easier.

2. By including the argument declaration, we announced to other programmers what they are expected to pass to our function. We have made it easier to understand our function.

   We have also told Mata what to expect and, if some other program attempts to use our function incorrectly, Mata will stop execution.

   Just as in (1), we have made debugging easier.

3. By including the inside declaration, we have told Mata what variables we will need and how we will be using them. Mata can do two things with that information: first, it can make sure that we are using the variables correctly (making debugging easier again), and second, Mata can produce more efficient code (making our function run faster).

Interactively, we admit that we sometimes define functions without declarations. For more careful work, however, we include them.

Types, element types, and organizational types

When you use Mata interactively, you just willy-nilly create new variables:

```
: n = 2
: A = (1,2 \ 3,4)
: z = (sqrt(-4+0i), sqrt(4))
```

When you create a variable, you may not think about the type, but Mata does. n above is, to Mata, a real scalar. A is a real matrix. z is a complex row vector.

Mata thinks of the type of a variable as having two parts:

1. the type of the elements the variable contains (such as real or complex) and

2. how those elements are organized (such as a row vector or a matrix).

We call those two types the *eltype*—element type—and *orgtype*—organizational type. The *eltypes* and *orgtypes* are

eltype	*orgtype*
transmorphic	matrix
numeric	vector
real	rowvector
complex	colvector
string	scalar
pointer	

You may choose one of each and so describe all the types Mata understands.

Implicit declarations

When you do not declare an object, Mata behaves as if you declared it to be transmorphic matrix:

1. transmorphic means that the matrix can be real, complex, string, or pointer.

2. matrix means that the organization is to be $r \times c$, $r \geq 0$ and $c \geq 0$.

At one point in your function, a transmorphic matrix might be a real scalar (real being a special case of transmorphic and scalar being a special case of a matrix when $r = c = 1$), and at another point, it might be a string colvector (string being a special case of transmorphic, and colvector being a special case of a matrix when c = 1).

Consider our swaprows() function without declarations,

```
function swaprows(A, i1, i2)
{
    B = A
    v = B[i1, .]
    B[i1, .] = B[i2, .]
    B[i2, .] = v
    return(B)
}
```

The result of compiling this function is just as if the function read

```
transmorphic matrix swaprows(transmorphic matrix A,
                             transmorphic matrix i1,
                             transmorphic matrix i2)
{
    transmorphic matrix      B
    transmorphic matrix      v
    B = A
    v = B[i1, .]
    B[i1, .] = B[i2, .]
    B[i2, .] = v
    return(B)
}
```

When we declare a variable, we put restrictions on it.

Element types

There are six *eltypes*, or element types:

1. transmorphic, which means real, complex, string, or pointer.

2. numeric, which means real or complex.

3. real, which means that the elements are real numbers, such as 1, 3, −50, and 3.14159.

4. complex, which means that each element is a pair of numbers, which are given the interpretation $a + bi$. complex is a storage type; the number stored in a complex might be real, such as $2 + 0i$.

5. string, which means the elements are strings of text. Each element may contain up to 2,147,483,647 characters and strings may (need not) contain binary 0; i.e., strings may be binary strings or text strings.

6. pointer means the elements are pointers to (addresses of) other Mata matrices, vectors, scalars, or even functions; see [M-2] **pointers**.

Organizational types

There are five *orgtypes*, or organizational types:

    1. matrix, which means $r \times c$, $r \geq 0$ and $c \geq 0$.

    2. vector, which means $1 \times n$ or $n \times 1$, $n \geq 0$.

    3. rowvector, which means $1 \times n$, $n \geq 0$.

    4. colvector, which means $n \times 1$, $n \geq 0$.

    5. scalar, which means 1×1.

Sharp-eyed readers will note that vectors and matrices can have zero rows or columns! See [M-2] **void** for more information.

Function declarations

Function declarations are the declarations that appear in front of the function name, such as

```
real matrix swaprows(...)
{
    ...
}
```

The syntax for what may appear there is

```
function
```
type $\big[\,$function$\,\big]$

void $\big[\,$function$\,\big]$

Something must appear in front of the name, and if you do not want to declare the type (which makes the type transmorphic matrix), you just put the word function:

```
function swaprows(...)
{
    ...
}
```

You may also declare the type and include the word function if you wish,

```
real matrix function swaprows(...)
{
    ...
}
```

but most programmers omit the word function; it makes no difference.

In addition to all the usual types, void is a type allowed only with functions—it states that the function returns nothing:

```
void _swaprows(real matrix A, real scalar i1, real scalar i2)
{
        real rowvector  v
        v = A[i1, .]
        A[i1, .] = A[i2, .]
        A[i2, .] = v
}
```

The function above returns nothing; it instead modifies the matrix it is passed. That might be useful to save memory, especially if every use of the original swaprows() was going to be

```
A = swaprows(A, i1, i2)
```

In any case, we named this new function _swaprows() (note the underscore), to flag the user that there is something odd and deserving caution concerning the use of this function.

void, that is to say, returning nothing, is also considered a special case of a transmorphic matrix because Mata secretly returns a 0×0 real matrix, which the caller just discards.

Argument declarations

Argument declarations are the declarations that appear inside the parentheses, such as

```
... swaprows(real matrix A, real scalar i1, real scalar i2)
{
        ...
}
```

The syntax for what may appear there is

$$\left[\,type\,\right]\ argname\ \left[\,,\ \left[\,type\,\right]\ argname\ \left[\,,\ \dots\,\right]\right]$$

The names are required—they specify how we will refer to the argument—and the types are optional. Omit the type and transmorphic matrix is assumed. Specify the type, and it will be checked when your function is called. If the caller attempts to use your function incorrectly, Mata will stop the execution and complain.

The by-address calling convention

Arguments are passed to functions by address, not by value. If you change the value of an argument, you will change the caller's argument. That is what made _swaprows() (above) work. The caller passed us A and we changed it. And that is why in the original version of swaprows(), the first line read

```
B = A
```

we did our work on B, and returned B. We did not want to modify the caller's original matrix.

You do not ordinarily have to make copies of the caller's arguments, but you do have to be careful if you do not want to change the argument. That is why in all the official functions (with the single exception of st_view()—see [M-5] **st_view( )**), if a function changes the caller's argument, the function's name starts with an underscore. The reverse logic does not hold: some functions start with an underscore and do not change the caller's argument. The underscore signifies caution, and you need to read the function's documentation to find out what it is you need to be cautious about.

Variable declarations

The variable declarations are the declarations that appear at the top of the body of a function:

```
... swaprows(...)
{
        real matrix       B
        real rowvector    v

        ...
}
```

These declarations are optional. If you omit them, Mata will observe that you are using B and v in your code, and then Mata will compile your code just as if you had declared the variables to be `transmorphic matrix`, meaning that the resulting compiled code might be a little more inefficient than it could be, but that is all.

The variable declarations are optional as long as you have not `mata set matastrict on`; see [M-3] **mata set**. Some programmers believe so strongly that variables really ought to be declared that Mata provides a provision to issue an error when they forget.

In any case, these declarations—explicit or implicit—define the variables we will use. The variables we use in our function are private—it does not matter if there are other variables named B and v floating around somewhere. Private variables are created when a function is invoked and destroyed when the function ends. The variables are private but, as explained above, if we pass our variables to another function, that function may change their values. Most functions do not.

The syntax for declaring variables is

> *type*   *varname* $\big[$, *varname* $\big[$, ... $\big]\big]$
>
> external $\big[$*type*$\big]$ *varname* $\big[$, *varname* $\big[$, ... $\big]\big]$

`real matrix B` and `real rowvector v` match the first syntax.

Linking to external globals

The second syntax has to do with linking to global variables. When you use Mata interactively and type

```
: n = 2
```

you create a variable named n. That variable is global. When you code inside a function

```
... myfunction(...)
{
        external n

        ...
}
```

The n variable your function will use is the global variable named n. If your function were to examine the value of n right now, it would discover that it contained 2.

If the variable does not already exist, the statement `external n` will create it. Pretend that we had not previously defined n. If `myfunction()` were to examine the contents of n, it would discover that n is a 0×0 matrix. That is because we coded

```
external n
```

and Mata behaved as if we had coded

```
external transmorphic matrix n
```

Let's modify myfunction() to read:

```
... myfunction(...)
{
        external real scalar n

        ...
}
```

Let's consider the possibilities:

1. n does not exist. Here external real scalar n will create n—as a real scalar, of course—and set its value to missing.

    If n had been declared a rowvector, a 1×0 vector would have been created.

    If n had been declared a colvector, a 0×1 vector would have been created.

    If n had been declared a vector, a 0×1 vector would have been created. Mata could just as well have created a 1×0 vector, but it creates a 0×1.

    If n had been declared a matrix, a 0×0 matrix would have been created.

2. n exists, and it is a real scalar. Our function executes, using the global n.

3. n exists, and it is a real 1×1 rowvector, colvector, or matrix. The important thing is that it is 1×1; our function executes, using the global n.

4. n exists, but it is complex or string or pointer, or it is real but not 1×1. Mata issues an error message and aborts execution of our function.

Complicated systems of programs sometimes find it convenient to communicate via globals. Because globals are globals, we recommend that you give your globals long names. A good approach is to put the name of your system as a prefix:

```
... myfunction(...)
{
        external real scalar mysystem_n

        ...
}
```

For another approach to globals, see [M-5] **findexternal( )** and [M-5] **valofexternal( )**.

Also See

[M-2] **intro** — Language definition

Title

[M-2] do — do ... while (exp)

Syntax

> do *stmt* while (*exp*)

> do {
>   *stmts*
> } while (*exp*)

where *exp* must evaluate to a real scalar.

Description

do executes *stmt* or *stmts* one or more times, until *exp* is zero.

Remarks

One common use of do is to loop until convergence:

```
do {
    lasta = a
    a = get_new_a(lasta)
} while (mreldif(a, lasta)>1e-10)
```

The loop is executed at least once, and the conditioning expression is not executed until the loop's body has been executed.

Also See

[M-2] **for** — for (exp1; exp2; exp3) stmt

[M-2] **while** — while (exp) stmt

[M-2] **break** — Break out of for, while, or do loop

[M-2] **continue** — Continue with next iteration of for, while, or do loop

[M-2] **intro** — Language definition

Title

[M-2] errors — Error codes

Description

When an error occurs, Mata presents a number as well as text describing the problem. The codes are presented below.

Also the error codes can be used as an argument with _error(), see [M-5] **error( )**.

Mata's error codes are a special case of Stata's return codes. In particular, they are the return codes in the range 3000–3999. In addition to the 3000-level codes, it is possible for Mata functions to generate any Stata error message and return code.

Remarks

Error messages in Mata break into two classes: errors that occur when the function is compiled (code 3000) and errors that occur when the function is executed (codes 3001–3999).

Compile-time error messages look like this:

```
: 2,,3
invalid expression
r(3000);

: "this" + "that
mismatched quotes
r(3000);
```

The text of the message varies according to the error made, but the error code is always 3000.

Run-time errors look like this:

```
: myfunction(2,3)
              solve():  3200  conformability error
              mysub():     -  function returned error
          myfunction():    -  function returned error
              <istmt>:     -  function returned error
r(3200);
```

The output is called a traceback log. Read from bottom to top, it says that what we typed (the <istmt>) called myfunction(), which called mysub(), which called solve() and, at that point, things went wrong. The error is possibly in solve(), but since solve() is a library function, it is more likely that the error is in how mysub() called solve(). Moreover, the error is seldom in the program listed at the top of the traceback log because the log lists the identity of the program that detected the error. Say solve() did have an error. Then the traceback log would probably have read something like

```
                   *:  3200  conformability error
              solve():     -  function returned error
              mysub():     -  function returned error
          myfunction():    -  function returned error
              <istmt>:     -  function returned error
```

The above log says the problem was detected by ∗ (the multiplication operator), and at that point, solve() would be suspect, because one must ask, why did solve() use the multiplication operator incorrectly?

In any case, let's assume that the problem is not with solve(). Then you would guess the problem lies with mysub(). If you have used mysub() in many previous programs without problems, however, you might now shift your suspicion to myfunction(). If myfunction() is always trustworthy, perhaps you should not have typed myfunction(2,3). That is, perhaps you are using myfunction() incorrectly.

The error codes

3000. (message varies)
    There is an error in what you have typed. Mata cannot interpret what you mean.

3001. incorrect number of arguments
    The function expected, say, three arguments and received two, or five. Or the function allows between three and five arguments, but you supplied too many or too few. Fix the calling program.

3002. identical arguments not allowed
    You have called a function specifying the same variable more than once. Usually this would not be a problem, but here, it is, usually because the supplied arguments are matrices that the function wishes to overwrite with different results. For instance, say function $f(A, B, C)$ examines matrix A and returns a calculation based on A in B and C. The function might well complain that you specified the same matrix for B and C.

3010. attempt to dereference NULL pointer
    The program made reference to ∗s, and s contains NULL; see [M-2] **pointers**.

3011. invalid lval
    In an assignment, what appears on the left-hand side of the equals is not something to which a value can be assigned; see [M-2] **op_assignment**.

3012. undefined operation on a pointer
    You have, for instance, attempted to add two pointers; see [M-2] **pointers**.

3101. matrix found where function required
    A particular argument to a function is required to be a function, and a matrix was found instead.

3102. function found where matrix required
    A particular argument to a function is required to be a matrix, vector, or scalar, and a function was found instead.

3103. view found where array required
    In general, view matrices can be used wherever a matrix is required, but there are a few exceptions, both in low-level routines and in routines that wish to write results back to the argument. Here a view is not acceptable. If V is the view variable, simply code $X = V$ and then pass X in its stead. See [M-5] **st_view()**.

3104. array found where view required
    A function argument was specified with a matrix that was not a view, and a view was required. See [M-5] **st_view()**.

3120. `attempt to dereference NULL pointer`
A pointer was equal to NULL and you put an * in front of it; see [M-2] **pointers**.

3200. `conformability error`
A matrix, vector, or scalar has the wrong number of rows and/or columns for what is required. Adding a 2×3 matrix to a 1×4 would result in this error.

3201. `vector required`
An argument is required to be $r \times 1$ or $1 \times c$, and a matrix was found instead.

3202. `rowvector required`
An argument is required to be $1 \times c$ and it is not.

3203. `colvector required`
An argument is required to be $r \times 1$ and it is not.

3204. `matrix found where scalar required`
An argument is required to be 1×1 and it is not.

3205. `square matrix required`
An argument is required to be $n \times n$ and it is not.

3250. `type mismatch`
The *eltype* of an argument does not match what is required. For instance, perhaps a real was expected and a string was received. See *eltype* in [M-6] **glossary**.

3251. `nonnumeric found where numeric required`
An argument was expected to be real or complex and it is not.

3252. `noncomplex found where complex required`
An argument was expected to be complex and it is not.

3253. `nonreal found where real required`
An argument was expected to be real and it is not.

3254. `nonstring found where string required`
An argument was expected to be string and it is not.

3255. `real or string required`
An argument was expected to be real or string and it is not.

3256. `numeric or string required`
An argument was expected to be real, complex, or string and it is not.

3257. `nonpointer found where pointer required`
An argument was expected to be a pointer and it is not.

3258. `nonvoid found where void required`
An argument was expected to be void and it is not.

3300. `argument out of range`
The *eltype* and *orgtype* of the argument are correct, but it contains an invalid value, such as you asking for the 20th row of a 4×5 matrix. See *eltype* and *orgtype* in [M-6] **glossary**.

3301. `subscript invalid`
The subscript is out of range (refers to a row or column that does not exist) or contains the wrong number of elements. See [M-2] **subscripts**.

3302. invalid %fmt
The %fmt for formatting data is invalid. See [M-5] **printf( )** and see [U] **12.5 Formats: controlling how data are displayed**.

3303. invalid permutation vector
The vector specified does not meet the requirements of a permutation vector, namely, that an n-element vector contain a permutation of the integers 1 through n. See [M-1] **permutation**.

3351. argument has missing values
In general, Mata is tolerant of missing values, but there are exceptions. This function does not allow the matrix, vector, or scalar to have missing values.

3352. singular matrix
The matrix is singular and the requested result cannot be carried out. If singular matrices are a possibility, then you are probably using the wrong function.

3353. matrix not positive definite
The matrix is non–positive definite and the requested results cannot be computed. If non–positive definite matrices are a possibility, then you are probably using the wrong function.

3360. failure to converge
The function that issued this message used an algorithm that the function expected would converge but did not, probably because the input matrix was extreme in some way.

3492. resulting string too long
A string the function was attempting to produce became too long. Since the maximum length of strings in Mata is 2,147,483,647 characters, it is unlikely that Mata imposed the limit. Review the documentation on the function for the source of the limit that was imposed (e.g., perhaps a string was being produced for use by Stata). In any case, this error does not arise because of an out-of-memory situation. It arises because some limit was imposed.

3498. (message varies)
An error specific to this function arose. The text of the message should describe the problem.

3499. _____ not found
The specified variable or function could not be found. For a function, it was not already loaded, it is not in the libraries, and there is no .mo file with its name.

3500. invalid Stata variable name
A variable name—which name is contained in a Mata string variable—is not appropriate for use with Stata.

3598. Stata returned error
You are using a Stata interface function and have asked Stata to perform a task. Stata could not or refused.

3601. invalid file handle
The number specified does not correspond to an open file handle; see [M-5] **fopen( )**.

3602. invalid filename
The filename specified is invalid.

3603. invalid file mode
The file mode (whether read, write, read-write, etc.) specified is invalid; see [M-5] **fopen( )**.

3611. `too many open files`
The maximum number of files that may be open simultaneously is 50, although your operating system may not allow that many.

3621. `attempt to write read-only file`
The file was opened read-only and an attempt was made to write into it.

3622. `attempt to read write-only file`
The file was opened write-only and an attempt was made to read it.

3623. `attempt to seek append-only file`
The file was opened append-only and then an attempt was made to seek into the file; see [M-5] **fopen( )**.

3698. `file seek error`
An attempt was made to seek to an invalid part of the file, or the seek failed for other reasons; see [M-5] **fopen( )**.

3900. `out of memory`
Mata is out of memory; the operating system refused to supply what Mata requested. There is no Mata or Stata setting that affects this, and so nothing in Mata or Stata to reset in order to get more memory. You must take up the problem with your operating system.

3901. `macro memory in use`
This error message should not occur; please notify StataCorp if it does.

3930. `error in LAPACK routine`
The linear-algebra LAPACK routines—see [M-1] **LAPACK**—generated an error that Mata did not expect. Please notify StataCorp if you should receive this error.

3995. `unallocated function`
This error message should not occur; please notify StataCorp if it does.

3996. `built-in unallocated`
This error message should not occur; please notify StataCorp if it does.

3997. `unimplemented opcode`
This error message should not occur; please notify StataCorp if it does.

3998. `stack overflow`
Your program nested too deeply. For instance, imagine calculating the factorial of n by recursively calling yourself and then requesting the factorial of 1e+100. Functions that call themselves in an infinite loop inevitably cause this error.

Also See

[M-5] **error( )** — Issue error message

[M-2] **intro** — Language definition

Title

Syntax

exp

Description

exp is used in syntax diagrams to mean "any valid expression may appear here". Expressions can range from being simple constants

```
2
"this"
3+2i
```

to being names of variables

```
A
beta
varwithverylongname
```

to being a full-fledged scalar, string, or matrix expression:

```
sqrt(2)/2
substr(userinput, 15, strlen(otherstr))
conj(X)'X
```

Remarks

Remarks are presented under the following headings:

> *What's an expression*
> *Assignment suppresses display, as does (void)*
> *The pieces of an expression*
> *Numeric literals*
> *String literals*
> *Variable names*
> *Operators*
> *Functions*

What's an expression

Everybody knows what an expression is: expressions are things like 2+3 and invsym(X'X)*X'y. Simpler things are also expressions, such as numeric constants

2   is an expression

and string literals

"hi there"   is an expression

and function calls:

    `sqrt(2)`   is an expression

Even when functions do not return anything (the function is void), the code that causes the function to run is an expression. For instance, the function `swap()` (see [M-5] **swap( )**) interchanges the contents of its arguments and returns nothing. Even so,

    `swap(A, B)`   is an expression

Assignment suppresses display, as does (void)

The equals sign assigns the result of an expression to a variable. For instance,

    `a = 2 + 3`

assigns 5 to a. When the result of an expression is not assigned to a variable, the result is displayed at the terminal. This is true of expressions entered interactively and of expressions coded in programs. For instance, given the program

```
function example(a, b)
{
        "the answer is"
        a+b
}
```

executing `example()` produces

```
: example(2, 3)
  the answer is
  5
```

The fact that 5 appeared is easy enough to understand; we coded the expression a+b without assigning it to another variable. The fact that "the answer is" also appeared may surprise you. Nevertheless, we coded `"the answer is"` in our program, and that is an example of an expression, and since we did not assign the expression to a variable, it was displayed.

In programming situations, there will be times when you want to execute a function—call it `setup()`— but do not care what the function returns, even though the function itself is not void (i.e., it returns something). If you code

```
function example(...)
{
        ...
        setup(...)
        ...
}
```

the result will be to display what `setup()` returns. You have two alternatives. You could assign the result of `setup` to a variable even though you will subsequently not use the variable

```
function example(...)
{
        ...
        result = setup(...)
        ...
}
```

or you could cast the result of the function to be void:

```
function example(...)
{
        ...
        (void) setup(...)
        ...
}
```

Placing (void) in front of an expression prevents the result from being displayed.

The pieces of an expression

Expressions comprise

    numeric literals
    string literals
    variable names
    operators
    functions

Numeric literals

Numeric literals are just numbers

```
2
3.14159
-7.2
5i
1.213e+32
1.213E+32
1.921fb54442d18X+001
1.921fb54442d18x+001

.
.a
.b
```

but you can suffix an i onto the end to mean imaginary, such as 5i above. To create complex numbers, you combine real and imaginary numbers using the + operator, as in 2+5i. In any case, you can put the i on the end of any literal, so 1.213e+32i is valid, as is 1.921fb54442d18X+001i.

1.921fb54442d18X+001i is a formidable-looking beast, with or without the i. 1.921fb54442d18X+001 is a way of writing floating-point numbers in binary; it is described in [U] **12.5 Formats: controlling how data are displayed**. Most people never use it.

Also, numeric literals include Stata's missing values, ., .a, .b, ..., .z.

Complex variables may contain missing just as real variables may, but they get only one: .a+.bi is not allowed. A complex variable contains a valid complex value, or it contains ., .a, .b, ..., .z.

String literals

String literals are enclosed in double quotes or in compound double quotes:

```
"the answer is"
"a string"
'"also a string"'
'"The "factor" of a matrix"'
""
'""'
```

Strings in Mata contain between 0 and 2,147,483,647 characters. "" or '""' is how one writes the 0-length string.

Any ASCII character may appear in the string, but no provision is provided for typing the unprintable ASCII characters into the string literal. Instead, you use the char() function; see [M-5] **ascii( )**. For instance, char(13) is carriage return, so the expression

```
"my string" + char(13)
```

produces "my string" followed by a carriage return.

No character is given a special interpretation. In particular, backslash (\) is given no special meaning by Mata. The string literal "my string\n" is just that: the characters "my string" followed by a backslash followed by an "n". Some functions, such as printf() (see [M-5] **printf( )**), give a special meaning to the two-character sequence \n, but that special interpretation is a property of the function, not Mata, and is noted in the function's documentation.

Strings are not zero (null) terminated in Mata. Mata knows that the string "hello" is of length 5, but it does not achieve that knowledge by padding a binary 0 as the string's fifth character. Thus strings may be used to hold binary information.

Although Mata gives no special interpretation to binary 0, some Mata functions do. For instance, strmatch(s, pattern) returns 1 if s matches pattern and 0 otherwise; see [M-5] **strmatch( )**. For this function, both strings are considered to end at the point they contain a binary 0, if they contain a binary 0. Most strings do not, and then the function considers the entire string. In any case, if there is special treatment of binary 0, that is on a function-by-function basis, and a note of that is made in the function's documentation.

Variable names

Variable names are just that. Names are case sensitive and no abbreviations are allowed:

```
X
x
MyVar
VeryLongVariableNameForUseInMata
MyVariable
```

The maximum length of a variable name is 32 characters.

Operators

Operators, listed by precedence, low to high

Operator	Operator name	Documentation
$a = b$	assignment	[M-2] **op_assignment**
$a\ ?\ b\ :\ c$	conditional	[M-2] **op_conditional**
$a \setminus b$	column join	[M-2] **op_join**
$a :: b$	column to	[M-2] **op_range**
$a\ ,\ b$	row join	[M-2] **op_join**
$a\ ..\ b$	row to	[M-2] **op_range**
$a :| b$	e.w. or	[M-2] **op_colon**
$a\ |\ b$	or	[M-2] **op_logical**
$a :\& b$	e.w. and	[M-2] **op_colon**
$a\ \&\ b$	and	[M-2] **op_logical**
$a :== b$	e.w. equal	[M-2] **op_colon**
$a == b$	equal	[M-2] **op_logical**
$a :>= b$	e.w. greater than or equal	[M-2] **op_colon**
$a >= b$	greater than or equal	[M-2] **op_logical**
$a :<= b$	e.w. less than or equal	[M-2] **op_colon**
$a <= b$	less than or equal	[M-2] **op_logical**
$a :< b$	e.w. less than	[M-2] **op_colon**
$a < b$	less than	[M-2] **op_logical**
$a :> b$	e.w. greater than	[M-2] **op_colon**
$a > b$	greater than	[M-2] **op_logical**
$a :!= b$	e.w. not equal	[M-2] **op_colon**
$a != b$	not equal	[M-2] **op_logical**
$a :+ b$	e.w. addition	[M-2] **op_colon**
$a + b$	addition	[M-2] **op_arith**
$a :- b$	e.w. subtraction	[M-2] **op_colon**
$a - b$	subtraction	[M-2] **op_arith**
$a :* b$	e.w. multiplication	[M-2] **op_colon**
$a * b$	multiplication	[M-2] **op_arith**
$a \# b$	Kronecker	[M-2] **op_kronecker**
$a :/ b$	e.w. division	[M-2] **op_colon**
a / b	division	[M-2] **op_arith**
$-a$	negation	[M-2] **op_arith**
$a :\hat{\ } b$	e.w. power	[M-2] **op_colon**
$a\ \hat{\ }\ b$	power	[M-2] **op_arith**
a'	transposition	[M-2] **op_transpose**
$*a$	contents of	[M-2] **pointers**
$\&a$	address of	[M-2] **pointers**
$!a$	not	[M-2] **op_logical**
$a[exp]$	subscript	[M-2] **subscripts**
$a[|exp|]$	range subscript	[M-2] **subscripts**
$a++$	increment	[M-2] **op_increment**
$a--$	decrement	[M-2] **op_increment**
$++a$	increment	[M-2] **op_increment**
$--a$	decrement	[M-2] **op_increment**

(e.w. = elementwise)

Functions

Functions supplied with Mata are documented in [M-5]. An index to the functions can be found in [M-4] **intro**.

Reference

Gould, W. 2006. Mata Matters: Precision. *Stata Journal* 6: 550–560.

Also See

[M-2] **intro** — Language definition

Title

Syntax

for (*exp*$_1$; *exp*$_2$; *exp*$_3$) *stmt*

for (*exp*$_1$; *exp*$_2$; *exp*$_3$) {
    stmts
}

where *exp*$_1$ and *exp*$_3$ are optional, and *exp*$_2$ must evaluate to a real scalar.

Description

for is equivalent to

exp$_1$
while (*exp*$_2$) {
    stmt(s)
    exp$_3$
}

stmt(s) is executed zero or more times. The loop continues as long as *exp2* is not equal to zero.

Remarks

To understand for, enter the following program

```
function example(n)
{
    for (i=1; i<=n; i++) {
        printf("i=%g\n", i)
    }
    printf("done\n")
}
```

and run example(3), example(2), example(1), example(0), and example(-1).

Common uses of for include

```
for (i=1; i<=rows(A); i++) {
    for (j=1; j<=cols(A); j++) {
        ...
    }
}
```

Also See

[M-2] **semicolons** — Use of semicolons

[M-2] **do** — do . . . while (exp)

[M-2] **while** — while (exp) stmt

[M-2] **break** — Break out of for, while, or do loop

[M-2] **continue** — Continue with next iteration of for, while, or do loop

[M-2] **intro** — Language definition

Title

[M-2] ftof — Passing functions to functions

Syntax

> *example*(..., &*somefunction*(), ...)

where *example*() is coded

```
function example(..., f, ...)
{
        ...
        (*f)(...)
        ...
}
```

Description

Functions can receive other functions as arguments.

Below is described (1) how to call a function that receives a function as an argument and (2) how to write a function that receives a function as an argument.

Remarks

Remarks are presented under the following headings:

> Passing functions to functions
> Writing functions that receive functions, the simplified convention
> Passing built-in functions

Passing functions to functions

Someone has written a program that receives a function as an argument. We will imagine that function is

> *real scalar* deriv(*function*(), *x*)

and that deriv() numerically evaluates the derivative of *function*() at *x*. The documentation for deriv() tells you to write a function that takes one argument and returns the evaluation of the function at that argument, such as

```
real scalar expratio(real scalar x)
{
        return(exp(x)/exp(-x))
}
```

To call deriv() and have it evaluate the derivative of expratio() at 3, you code

```
deriv(&expratio(), 3)
```

To pass a function to a function, you code & in front of the function's name and () after. Coding &expratio() passes the address of the function expratio() to deriv().

Writing functions that receive functions, the simplified convention

To receive a function, you include a variable among the program arguments to receive the function—we will use *f*—and you then code (*f)(...) to call the passed function. The code for deriv() might read

```
function deriv(f, x)
{
        return( ((*f)(x+1e-6) - (*f)(x)) / 1e-6 )
}
```

or, if you prefer to be explicit about your declarations,

```
real scalar deriv(pointer scalar f, real scalar x)
{
        return( ((*f)(x+1e-6) - (*f)(x)) / 1e-6 )
}
```

or, if you prefer to be even more explicit:

```
real scalar deriv(pointer(real scalar function) scalar f,
                  real scalar x)
{
        return( ((*f)(x+1e-6) - (*f)(x)) / 1e-6 )
}
```

In any case, using pointers, you type (*f)(...) to execute the function passed. See [M-2] **pointers** for more information.

Aside: the function deriv() would work but, because of the formula it uses, would return very inaccurate results.

Passing built-in functions

You cannot pass built-in functions to other functions. For instance, [M-5] **exp( )** is built in, which is revealed by [M-3] **mata which**:

```
: mata which exp()
  exp():  built-in
```

Not all official functions are built in. Many are implemented in Mata as library functions, but exp() is built in and coding &exp() will result in an error. If you wanted to pass exp() to a function, create your own version of it

```
: function myexp(x) return(exp(x))
```

and then pass &myexp().

Also See

[M-2] **intro** — Language definition

Title

[M-2] **goto** — goto label

Syntax

> *label*:   ...
>
>   ...
>
>   goto *label*

where *label*: may occur before or after the goto.

Description

goto *label* causes control to pass to the statement following *label*:. *label* may be any name up to eight characters long.

Remarks

These days, good style is to avoid using goto.

goto is useful when translating a FORTRAN program, such as

```
        A = 4.0e0/3.0e0
10 B = A - 1.0e0
        C = B + B + B
        EPS = DABS(C - 1.0e0)
        if (EPS.EQ.0.0e0) GOTO 10
```

The Mata translation is

```
            a = 4/3
s10:    b = a - 1
            c = b + b + b
            eps = abs(c-1)
            if (eps==0) goto s10
```

although

```
a = 4/3
do {
        b = a - 1
        c = b + b + b
        eps = abs(c - 1)
} while (eps==0)
```

is more readable.

Reference

Gould, W. 2005. Mata Matters: Translating Fortran. *Stata Journal* 5: 421–441.

Also See

[M-2] **do** — do . . . while (exp)

[M-2] **for** — for (exp1; exp2; exp3) stmt

[M-2] **while** — while (exp) stmt

[M-2] **break** — Break out of for, while, or do loop

[M-2] **continue** — Continue with next iteration of for, while, or do loop

[M-2] **intro** — Language definition

Title

[M-2] **if** — if (exp) ... else ...

Syntax

```
if (exp) stmt1


if (exp) stmt1
else stmt2


if (exp) {
        stmts1
}
else {
        stmts2
}
if (exp1) ...
else if (exp2) ...
else if (exp3) ...
...
else ...
```

where exp, exp_1, exp_2, exp_3, ..., must evaluate to real scalars.

Description

if evaluates the expression, and if it is true (evaluates to a nonzero number), if executes the statement or statement block that immediately follows it; otherwise, if skips the statement or block.

if ... else evaluates the expression, and if it is true (evaluates to a nonzero number), if executes the statement or statement block that immediately follows it and skips the statement or statement block following the else; otherwise, it skips the statement or statement block immediately following it and executes the statement or statement block following the else.

Remarks

if followed by multiple elses is interpreted as being nested, i.e.,

```
if (exp₁) ...
else if (exp₂) ...
else if (exp₃) ...
...
else ...
```

is equivalent to

```
if (exp₁) ...
else {
        if (exp₂) ...
        else {
                if (exp₃) ...
                else {
                        ...
                }
        }
}
```

Also See

[M-2] **intro** — Language definition

Title

Syntax

$a + b$	addition
$a - b$	subtraction
$a * b$	multiplication
a / b	division
$a \char94 b$	power
$-a$	negation

where a and b may be numeric scalars, vectors, or matrices.

Description

The above operators perform basic arithmetic.

Remarks

Also see [M-2] **op_colon** for the :+, :−, :*, and :/ operators. Colon operators have relaxed conformability restrictions.

The * and :* multiplication operators can also perform string duplication—3*"a" = "aaa"—see [M-5] **strdup( )**.

Conformability

$a + b$, $a - b$:

a:	$r \times c$
b:	$r \times c$
result:	$r \times c$

$a * b$:

a:	$k \times n$	$k \times n$	1×1
b:	$n \times m$	1×1	$n \times m$
result:	$k \times m$	$k \times n$	$n \times m$

a / b:

a:	$r \times c$
b:	1×1
result:	$r \times c$

$a \char94 b$:

a:	1×1
b:	1×1
result:	1×1

$-a$:

a:	$r \times c$	
result:	$r \times c$	

Diagnostics

All operators return missing when arguments are missing.

$a*b$ with *a*: $k \times 0$ and *b*: $0 \times m$ returns a $k \times m$ matrix of zeros.

a/b returns missing when $b{=}{=}0$ or when a/b would result in overflow.

$a\hat{\ }b$ returns a real when both *a* and *b* are real; thus, (-4)^.5 evaluates to missing, whereas (-4+0i)^.5 evaluates to 2i.

$a\hat{\ }b$ returns missing on overflow.

Also See

[M-2] **exp** — Expressions

[M-2] **intro** — Language definition

Title

[M-2] op_assignment — Assignment operator

Syntax

lval = *exp*

where *exp* is any valid expression and where *lval* is

name
name[*exp*]
name[*exp*, *exp*]
name[|*exp*|]

In pointer use (advanced), *name* may be

*lval
*(*lval*)
*(*lval*[*exp*])
*(*lval*[*exp*, *exp*])
*(*lval*[|*exp*|])

in addition to being a variable name.

Description

= assigns the evaluation of *exp* to *lval*.

Do not confuse the = assignment operator with the == equality operator. Coding

    x = y

assigns the value of y to x. Coding

    if (x==y) ...   (*note doubled equal signs*)

performs the action if the value of *x* is equal to the value of *y*. See [M-2] **op_logical** for a description of the == equality operator.

If the result of an expression is not assigned to a variable, then the result is displayed at the terminal; see [M-2] **exp**.

Remarks

Remarks are presented under the following headings:

Assignment suppresses display
The equal-assignment operator
lvals, what appears on the left-hand side
Row, column, and element lvals
Pointer lvals

Assignment suppresses display

When you interactively enter an expression or code an expression in a program without the equal-assignment operator, the result of the expression is displayed at the terminal:

```
: 2 + 3
  5
```

When you assign the expression to a variable, the result is not displayed:

```
: x = 2 + 3
```

The equal-assignment operator

Equals is an operator, so in addition to coding

```
a = 2 + 3
```

you can code

```
a = b = 2 + 3
```

or

```
y = x / (denominator = sqrt(a+b))
```

or even

```
y1 = y2 = x / (denominator = sqrt(sum=a+b))
```

This last is equivalent to

```
sum = a + b
denominator = sqrt(sum)
y2 = x / denominator
y1 = y2
```

Equals binds weakly, so

```
a = b = 2 + 3
```

is interpreted as

```
a = b = (2 + 3)
```

and not

```
a = (b=2) + 3
```

lvals, what appears on the left-hand side

What appears to the left of the equals is called an *lval*, short for left-hand-side value. It would make no sense, for instance, to code

```
sqrt(4) = 3
```

and, as a matter of fact, you are not allowed to code that because sqrt(4) is not an *lval*:

```
: sqrt(4) = 3
invalid lval
r(3000);
```

An *lval* is anything that can hold values. A scalar can hold values

```
a = 3
x = sqrt(4)
```

a matrix can hold values

```
A = (1, 2 \ 3, 4)
B = invsym(C)
```

a matrix row can hold values

```
A[1,.] = (7, 8)
```

a matrix column can hold values

```
A[.,2] = (9 \ 10)
```

and finally, a matrix element can hold a value

```
A[1,2] = 7
```

lvals are usually one of the above forms. The other forms have to do with pointer variables, which most programmers never use; they are discussed under *Pointer lvals* below.

Row, column, and element lvals

When you assign to a row, column, or element of a matrix,

```
A[1,.] = (7, 8)
A[.,2] = (9 \ 10)
A[1,2] = 7
```

the row, column, or element must already exist:

```
: A = (1, 2 \ 3, 4)
: A[3,4] = 4
            <istmt>:  3301  subscript invalid
r(3301);
```

This is usually not an issue because, by the time you are assigning to a row, column, or element, the matrix has already been created, but in the event you need to create it first, use the J() function; see [M-5] **J()**. The following code fragment creates a 3×4 matrix containing the sum of its indices:

```
A = J(3, 4, .)
for (i=1; i<=3; i++) {
    for (j=1; j<=4; j++) A[i,j] = i + j
}
```

Pointer lvals

In addition to the standard *lvals*

```
A = (1, 2 \ 3, 4)
A[1,.] = (7, 8)
A[.,2] = (9 \ 10)
A[1,2] = 7
```

pointer *lvals* are allowed. For instance,

```
*p = 3
```

stores 3 in the address pointed to by pointer scalar p.

```
(*q)[1,2] = 4
```

stores 4 in the (1,2) element of the address pointed to by pointer scalar q, whereas

```
*Q[1,2] = 4
```

stores 4 in the address pointed to by the (1,2) element of pointer matrix Q.

```
*Q[2,1][1,3] = 5
```

is equivalent to

```
*(Q[2,1])[1,3] = 5
```

and stores 5 in the (1,3) element of the address pointed to by the (2,1) element of pointer matrix Q.

Pointers to pointers, pointers to pointers to pointers, etc., are also allowed. For instance,

```
**r = 3
```

stores 3 in the address pointed to by the address pointed to by pointer scalar r, whereas

```
*((*(Q[1,2]))[2,1])[3,4] = 7
```

stores 7 in the (3,4) address pointed to by the (2,1) address pointed to by the (1,2) address of pointer matrix Q.

Conformability

$a = b$:

> *input*:
> > b: $r \times c$
> *output*:
> > a: $r \times c$

Diagnostics

$a = b$ aborts with error if there is insufficient memory to store a copy of b in a.

Also See

[M-5] **swap( )** — Interchange contents of variables

[M-2] **exp** — Expressions

[M-2] **intro** — Language definition

Title

Syntax

a	:+	b	addition
a	:-	b	subtraction
a	:*	b	multiplication
a	:/	b	division
a	:^	b	power
a	:==	b	equality
a	:!=	b	inequality
a	:>	b	greater than
a	:>=	b	greater than or equal to
a	:<	b	less than
a	:<=	b	less than or equal to
a	:&	b	and
a	:|	b	or

Description

Colon operators perform element-by-element operations.

Remarks

Remarks are presented under the following headings:

> *C-conformability: element by element*
> *Usefulness of colon logical operators*
> *Use parentheses*

C-conformability: element by element

The colon operators perform the indicated operation on each pair of elements of a and b. For instance,

$$\begin{bmatrix} c & d \\ f & g \\ h & i \end{bmatrix} \quad :* \quad \begin{bmatrix} j & k \\ l & m \\ n & o \end{bmatrix} = \begin{bmatrix} c*j & d*k \\ f*l & g*m \\ h*n & i*o \end{bmatrix}$$

Also colon operators have a relaxed definition of conformability:

$$\begin{bmatrix} c \\ f \\ g \end{bmatrix} \quad :* \quad \begin{bmatrix} j & k \\ l & m \\ n & o \end{bmatrix} = \begin{bmatrix} c*j & c*k \\ f*l & f*m \\ g*n & g*o \end{bmatrix}$$

$$\begin{bmatrix} c & d \\ f & g \\ h & i \end{bmatrix} \quad :* \quad \begin{bmatrix} j \\ l \\ n \end{bmatrix} = \begin{bmatrix} c*j & d*j \\ f*l & g*l \\ h*n & i*n \end{bmatrix}$$

110

$$[\,c \quad d\,] \; :\!* \; \begin{bmatrix} j & k \\ l & m \\ n & o \end{bmatrix} = \begin{bmatrix} c*j & d*k \\ c*l & d*m \\ c*n & d*o \end{bmatrix}$$

$$\begin{bmatrix} c & d \\ f & g \\ h & i \end{bmatrix} \; :\!* \; [\,l \quad m\,] = \begin{bmatrix} c*l & d*m \\ f*l & g*m \\ h*l & i*m \end{bmatrix}$$

$$c \; :\!* \; \begin{bmatrix} j & k \\ l & m \\ n & o \end{bmatrix} = \begin{bmatrix} c*j & c*k \\ c*l & c*m \\ c*n & c*o \end{bmatrix}$$

$$\begin{bmatrix} c & d \\ f & g \\ h & i \end{bmatrix} \; :\!* \; j = \begin{bmatrix} c*j & d*j \\ f*j & g*j \\ h*j & i*j \end{bmatrix}$$

The matrices above are said to be c-conformable; the c stands for colon. The matrices have the same number of rows and columns, or one or the other is a vector with the same number of rows or columns as the matrix, or one or the other is a scalar.

C-conformability is relaxed, but not everything is allowed. The following is an error:

$$(c \; d \; e) \; :\!* \; \begin{bmatrix} f \\ g \\ h \end{bmatrix}$$

Usefulness of colon logical operators

It is worth paying particular attention to the colon logical operators because they can produce pattern vectors and matrices. Consider the matrix

```
: x = (5, 0 \ 0, 2 \ 3, 8)
: x
        1    2

    1   5    0
    2   0    2
    3   3    8
```

Which elements of x contain 0?

```
: x:==0
        1    2

    1   0    1
    2   1    0
    3   0    0
```

How many zeros are there in x?

```
: sum(x:==0)
    2
```

Use parentheses

Because of their relaxed conformability requirements, colon operators are not associative even when the underlying operator is. For instance, you expect $(a+b)+c == a+(b+c)$, at least ignoring numerical roundoff error. Nevertheless, $(a:+b):+c == a:+(b:+c)$ does not necessarily hold. Consider what happens when

$$
\begin{array}{ll}
a: & 1 \times 4 \\
b: & 5 \times 1 \\
c: & 5 \times 4
\end{array}
$$

Then $(a:+b):+c$ is an error because $a:+b$ is not c-conformable.

Nevertheless, $a:+(b:+c)$ is not an error and in fact produces a 5×4 matrix because $b:+c$ is 5×4, which is c-conformable with a.

Conformability

$a : op \ b$:

$$
\begin{array}{ll}
a: & r_1 \times c_1 \\
b: & r_2 \times c_2, \quad a \text{ and } b \text{ c-conformable} \\
result: & \mathtt{max}(r_1, r_2) \times \mathtt{max}(c_1, c_2)
\end{array}
$$

Diagnostics

The colon operators return missing and abort with error under the same conditions that the underlying operator returns missing and aborts with error.

Also See

[M-2] **exp** — Expressions

[M-2] **intro** — Language definition

Title

Syntax

$$a \; ? \; b \; : \; c$$

where a must evaluate to a real scalar, and b and c may be of any type whatsoever.

Description

The conditional operator returns b if a is true (a is not equal to 0) and c otherwise.

Remarks

Conditional operators

```
dof = (k==0 ? n-1 : n-k)
```

are more compact than the if−else alternative

```
if (k==0) dof = n-1
else      dof = n-k
```

and they can be used as parts of expressions:

```
mse = ess/(k==0 ? n-1 : n-k)
```

Conformability

$a \; ? \; b \; : \; c$:

a:	1×1	
b:	$r_1 \times c_1$	
c:	$r_2 \times c_2$	
$result$:	$r_1 \times c_1$ or	$r_2 \times c_2$

Diagnostics

In $a \; ? \; b \; : \; c$, only the necessary parts are evaluated: a and b if a is true, or a and c if a is false. However, the ++ and −− operators are always evaluated:

```
( k==0 ? i++ : j++)
```

increments both i and j, regardless of the value of k.

Also See

[M-2] **exp** — Expressions

[M-2] **intro** — Language definition

Title

Syntax

++*i*	increment before
--*i*	decrement before
i++	increment after
i--	decrement after

where *i* must be a real scalar.

Description

++*i* and *i*++ increment *i*; they perform the operation *i*=*i*+1. ++*i* performs the operation before the evaluation of the expression in which it appears, whereas *i*++ performs the operation afterward.

--*i* and *i*-- decrement *i*; they perform the operation *i*=*i*-1. --*i* performs the operation before the evaluation of the expression in which is appears, whereas *i*-- performs the operation afterward.

Remarks

These operators are used in code, such as

```
x[i++] = 2
x[--i] = 3
for (i=0; i<100; i++) {
    ...
}
if (++n > 10) {
    ...
}
```

Where these expressions appear, results are as if the current value of i were substituted, and in addition, i is incremented, either before or after the expression is evaluated. For instance,

```
x[i++] = 2
```

is equivalent to

```
x[i] = 2 ; i = i + 1
```

and

```
x[++i] = 3
```

is equivalent to

```
i = i + 1 ; x[i] = 3
```

115

Coding

```
for (i=0; i<100; i++) {
    ...
}
```

or

```
for (i=0; i<100; ++i) {
    ...
}
```

is equivalent to

```
for (i=0; i<100; i=i+1) {
    ...
}
```

because it does not matter whether the incrementation is performed before or after the otherwise null expression.

```
if (++n > 10) {
    ...
}
```

is equivalent to

```
n = n + 1
if (n > 10) {
    ...
}
```

whereas

```
if (n++ > 10) {
    ...
}
```

is equivalent to

```
if (n > 10) {
      n = n + 1
    ...
}
else    n = n + 1
```

The ++ and -- operators may be used only with real scalars and are usually associated with indexing or counting. They result in fast and readable code.

Conformability

$++i$, $--i$, $i++$, and $i--$:

        i: 1×1
    result: 1×1

Diagnostics

++ and -- are allowed with real scalars only. That is, ++i or i++ is valid, assuming i is a real scalar, but x[i,j]++ is not valid.

++ and -- abort with error if applied to a variable that is not a real scalar.

++i, i++, --i, and i-- should be the only reference to i in the expression. Do not code, for instance,

```
x[i++] = y[i]
x[++i] = y[i]
x[i] = y[i++]
x[i] = y[++i]
```

The value of i in the above expressions is formally undefined; whatever is its value, you cannot depend on that value being obtained by earlier or later versions of the compiler. Instead code

```
i++ ; x[i] = y[i]
```

or code

```
x[i] = y[i] ; i++
```

according to the desired outcome.

It is, however, perfectly reasonable to code

```
x[i++] = y[j++]
```

That is, multiple ++ and -- operators may occur in the same expression; it is multiple references to the target of the ++ and -- that must be avoided.

Also See

[M-2] **exp** — Expressions

[M-2] **intro** — Language definition

Title

Syntax

a , b

$a \setminus b$

Description

, and $\setminus$ are Mata's row-join and column-join operators.

Remarks

Remarks are presented under the following headings:

Comma and backslash are operators
Comma as a separator
Warning about the misuse of comma and backslash operators

Comma and backslash are operators

That , and $\setminus$ are operators cannot be emphasized enough. When one types

```
: (1, 2 \ 3, 4)
       1   2

  1 |  1   2
  2 |  3   4
```

one is tempted to think, "Ah, comma and backslash are how you separate elements when you enter a matrix." If you think like that, you will not appreciate the power of , and $\setminus$.

, and $\setminus$ are operators in the same way that $*$ and $+$ are operators.

, is the operator that takes a $r \times c_1$ matrix and a $r \times c_2$ matrix, and returns a $r \times (c_1 + c_2)$ matrix.

$\setminus$ is the operator that takes a $r_1 \times c$ matrix and a $r_2 \times c$ matrix, and returns a $(r_1 + r_2) \times c$ matrix.

, and $\setminus$ may be used with scalars, vectors, or matrices:

```
: a = (1 \ 2)
: b = (3 \ 4)
: a, b
       1   2

  1 |  1   3
  2 |  2   4

: c = (1, 2)
```

```
: d = (3, 4)
: c \ d
        1   2

1   1   2
2   3   4
```

, binds more tightly than \, meaning that $e, f \setminus g, h$ is interpreted as $(e, f) \setminus (g, h)$. In this, , and \ are no different from * and + operators: * binds more tightly than + and $e*f + g*h$ is interpreted as $(e*f) + (g*h)$.

Just as it sometimes makes sense to type $e*(f + g)*h$, it can make sense to type $e, (f \setminus g), h$:

```
: e = 1 \ 2
: f = 5 \ 6
: g = 3
: h = 4
: e,(g\h),f
        1   2   3

1   1   3   5
2   2   4   6
```

Comma as a separator

, has a second meaning in Mata: it is the argument separator for functions. When you type

```
: myfunc(a, b)
```

the comma that appears inside the parentheses is not the comma row-join operator; it is the comma argument separator. If you wanted to call myfunc() with second argument equal to row vector (1,2), you must type

```
: myfunc(a, (1,2))
```

and not

```
: myfunc(a, 1, 2)
```

because otherwise Mata will think you are trying to pass three arguments to myfunc(). When you open another set of parentheses inside a function's argument list, comma reverts back to its usual row-join meaning.

Warning about the misuse of comma and backslash operators

Misuse or mere overuse of , and \ can substantially reduce the speed with which your code executes. Consider the actions Mata must take when you code, say,

$a \setminus b$

First, Mata must allocate a matrix or vector containing rows(a)+rows(b) rows, then it must copy a into the new matrix or vector, and then it must copy b. Nothing inefficient has happened yet, but now consider

$(a \setminus b) \setminus c$

Picking up where we left off, Mata must allocate a matrix or vector containing $rows(a)+rows(b)+rows(c)$ rows, then it must copy $(a \setminus b)$ into the new matrix or vector, and then it must copy c. Something inefficient just happened: a was copied twice!

Coding

$$res = (a \setminus b) \setminus c$$

is convenient, but execution would be quicker if we coded

$$res = J(rows(a)+rows(b)+rows(c), \ cols(a), \ .)$$
$$res[1,.] = a$$
$$res[2,.] = b$$
$$res[3,.] = c$$

We do not want to cause you concern where none is due. In general, you would not be able to measure the difference between the more efficient code and coding $res = (a \setminus b) \setminus c$. But as the number of row or column operators stack up, the combined result becomes more and more inefficient. Even that is not much of a concern. If the inefficient construction itself is buried in a loop, however, and that loop is executed thousands of times, the inefficiency can be become important.

With a little thought, you can always substitute predeclaration using $J()$ (see [M-5] **J( )**) and assignment via subscripting.

Conformability

a,b:

a:	$r \times c_1$
b:	$r \times c_2$
result:	$r \times (c_1 + c_2)$

$a \setminus b$:

a:	$r_1 \times c$
b:	$r_2 \times c$
result:	$(r_1 + r_2) \times c$

Diagnostics

, and $\setminus$ abort with error if a and b are not of the same broad type.

Also See

[M-2] **exp** — Expressions

[M-2] **intro** — Language definition

Title

[M-2] **op_kronecker** — Kronecker direct-product operator

Syntax

$A\#B$

where A and B may be real or complex.

Description

$A\#B$ returns the Kronecker direct product.

binds tightly: $X*A\#B*Y$ is interpreted as $X*(A\#B)*Y$.

Remarks

The Kronecker direct product is also known as the Kronecker product, the direct product, the tensor product, and the outer product.

The Kronecker product $A\#B$ is the matrix $||a_{ij}*B||$.

Conformability

$A\#B$:

A:	$r_1 \times c_1$
B:	$r_2 \times c_2$
result:	$r_1*r_2 \times c_1*c_2$

Diagnostics

None.

Leopold Kronecker (1823–1891) was born in Liegnitz, Prussia (now Legnica, Poland), to a well-off family. He attended the universities of Berlin, Bonn, and Breslau before completing a doctorate on the complex roots of unity. For several years, Kronecker devoted himself to business interests while working on mathematics in his spare time, publishing particularly in number theory, elliptic functions, and the theory of equations. He later started giving lectures at the university in Berlin, as was his right as a member of the Academy of Science. In 1883, he was appointed as the chair. Kronecker came to believe that mathematical arguments should involve only finite numbers and a finite number of operations, which led to increasing mathematical and personal disagreements with those who worked on irrational numbers or nonconstructive existence proofs.

Reference

James, I. 2002. *Remarkable Mathematicians: From Euler to von Neumann.* Cambridge: Cambridge University Press.

Also See

[M-2] **exp** — Expressions

[M-2] **intro** — Language definition

Title

Syntax

$a == b$	true if a equals b
$a\ != b$	true if a not equal to b
$a > \ \ b$	true if a greater than b
$a >= b$	true if a greater than or equal to b
$a < \ \ b$	true if a less than b
$a <= b$	true if a less than or equal to b
$!a$	logical negation; true if $a==0$ and false otherwise
$a\ \&\ b$	true if $a!=0$ and $b!=0$
$a\ \|\ b$	true if $a!=0$ or $b!=0$
$a\ \&\&\ b$	synonym for $a\ \&\ b$
$a\ \|\|\ b$	synonym for $a\ \|\ b$

Description

The operators above perform logical comparisons, and operator ! performs logical negation. All operators evaluate to 1 or 0, meaning true or false.

Remarks

The operators above work as you would expect when used with scalars, and the comparison operators and the not operator have been generalized for use with matrices.

$a==b$ evaluates to true if a and b are p-conformable, of the same type, and the corresponding elements are equal. Of the same type means a and b are both numeric, both strings, or both pointers. Thus it is not an error to ask if a 2×2 matrix is equal to a 4×1 vector or if a string variable is equal to a real variable; they are not. Also $a==b$ is declared to be true if a or b are p-conformable and the number of rows or columns is zero.

$a!=b$ is equivalent to $!(a==b)$. $a!=b$ evaluates to true when $a==b$ would evaluate to false and evaluates to true otherwise.

The remaining comparison operators >, >=, <, and <= work differently from == and != in that they require a and b be p-conformable; if they are not, they abort with error. They return true if the corresponding elements have the stated relationship, and return false otherwise. If a or b is complex, the comparison is made in terms of the length of the complex vector; for instance, $a>b$ is equivalent to abs(a)>abs(b), and so $-3>2+0i$ is true.

$!a$, when a is a scalar, evaluates to 0 if a is not equal to zero and 1 otherwise. Applied to a vector or matrix, the same operation is carried out, element by element: $!(-1,0,1,2,.)$ evaluates to $(0,1,0,0,0)$.

& and | (*and* and *or*) may be used with scalars only. Because so many people are familiar with programming in the C language, Mata provides && as a synonym for & and || as a synonym for |.

Use of logical operators with pointers

In a pointer expression, NULL is treated as false and all other pointer values (address values) are treated as true. Thus the following code is equivalent

```
pointer x                          pointer x
...                                ...
if (x) {                           if (x!=NULL) {
    ...                                ...
}                                  }
```

The logical operators $a==b$, $a!=b$, $a\&b$, and $a|b$ may be used with pointers.

Conformability

$a==b$, $a!=b$:

a:	$r_1 \times c_1$
b:	$r_2 \times c_2$
result:	1×1

$a>b$, $a>=b$, $a<b$, $a<=b$:

a:	$r \times c$
b:	$r \times c$
result:	1×1

$!a$:

a:	$r \times c$
result:	$r \times c$

$a\&b$, $a|b$:

a:	1×1
b:	1×1
result:	1×1

Diagnostics

$a==b$ and $a!=b$ cannot fail.

$a>b$, $a>=b$, $a<b$, $a<=b$ abort with error if a and b are not p-conformable, if a and b are not of the same general type (numeric and numeric or string and string), or if a or b are pointers.

$!a$ aborts with error if a is not real.

$a\&b$ and $a|b$ abort with error if a and b are not both real or not both pointers. If a and b are pointers, pointer value NULL is treated as false and all other pointer values are treated as true. In all cases, a real equal to 0 or 1 is returned.

Also See

[M-2] **exp** — Expressions

[M-2] **intro** — Language definition

Title

Syntax

$a..b$ row range

$a::b$ column range

Description

The range operators create vectors that count from a to b.

$a..b$ returns a row vector.

$a::b$ returns a column vector.

Remarks

$a..b$ and $a::b$ count from a up to but not exceeding b, incrementing by 1 if $b>=a$ and by -1 if $b<a$.

1..4 creates row vector (1,2,3,4).

1::4 creates column vector (1\2\3\4).

-1..-4 creates row vector (-1,-2,-3,-4).

-1::-4 creates column vector (-1\-2\-3\-4).

1.5..4.5 creates row vector (1.5, 2.5, 3.5, 4.5).

1.5::4.5 creates column vector (1.5\ 2.5\ 3.5\ 4.5).

1.5..4.4 creates row vector (1.5, 2.5, 3.5).

1.5::4.4 creates column vector (1.5\ 2.5\ 3.5).

-1.5..-4.4 creates row vector (-1.5, -2.5, -3.5).

-1.5::-4.4 creates column vector (-1.5\ -2.5\ -3.5).

1..1 and 1::1 both return (1).

Conformability

$a..b$

a:	1×1
b:	1×1
result:	$1 \times \text{trunc}(\text{abs}(b - a)) + 1$

$a::b$

a:	1×1
b:	1×1
result:	$\text{trunc}(\text{abs}(b - a)) + 1 \times 1$

Diagnostics

$a..b$ and $a::b$ return missing if $a\text{>=}.$ or $b\text{>=}.$

Also See

[M-2] **exp** — Expressions

[M-2] **intro** — Language definition

Title

[M-2] op_transpose — Conjugate transpose operator

Syntax

A'

Description

A' returns the transpose of A or, if A is complex, the conjugate transpose.

Remarks

The $'$ postfix operator may be used on any type of matrix or vector: real, complex, string, or pointer:

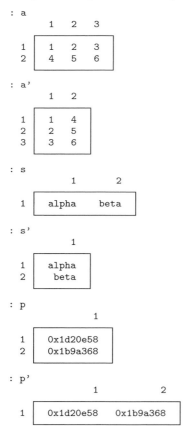

```
: z
               1           2

      1     1 + 2i     3 + 4i
      2     5 + 6i     7 + 8i


: z'
               1           2

      1     1 - 2i     5 - 6i
      2     3 - 4i     7 - 8i
```

When ' is applied to a complex, returned is the conjugate transpose. If you do not want this, code conj(z') or conj(z)'—it makes no difference; see [M-5] **conj( )**,

```
: conj(z')
               1           2

      1     1 + 2i     5 + 6i
      2     3 + 4i     7 + 8i
```

Or use the transposeonly() function; see [M-5] **transposeonly( )**:

```
: transposeonly(z)
               1           2

      1     1 + 2i     5 + 6i
      2     3 + 4i     7 + 8i
```

transposeonly() executes slightly faster than conj(z').

For real and complex A, also see [M-5] **_transpose( )**, which provides a way to transpose a matrix in place and so saves memory.

Conformability

A':
      A: $r \times c$
   result: $c \times r$

Diagnostics

The transpose operator cannot fail, but it is easy to use it incorrectly when working with complex quantities.

A user wanted to form $A*x$ but when he tried, got a conformability error. He thought x was a column vector, but it turned out to be a row vector, or perhaps it was the other way around. Anyway, he then coded $A*x'$, and the program worked and, even better, produced the correct answers. In his test, x was real.

Later, the user ran the program with complex x, and the program generated incorrect results, although it took him a while to notice. Study and study his code he did, before he thought about the innocuous $A*x'$. The transpose operator had not only changed x from being a row into being a column but had taken the conjugate of each element of x! He changed the code to read $A*$transposeonly(x).

The user no doubt wondered why the ′ transpose operator was not defined at the outset to be equivalent to transposeonly(). If it had been, then rather than telling the story of the man who was bitten by conjugate transpose when he only wanted the transpose, we would have told the story of the woman who was bitten by the transpose when she needed the conjugate transpose. There are, in fact, more of the latter stories than there are of the former.

Also See

[M-5] **conj( )** — Complex conjugate

[M-5] **_transpose( )** — Transposition in place

[M-5] **transposeonly( )** — Transposition without conjugation

[M-2] **exp** — Expressions

[M-2] **intro** — Language definition

Title

Syntax

> function *functionname*($|arg$ $[$, *arg* $[$, ... $]]$) { ...}

> function *functionname*(*arg*, $|arg$ $[$, ... $]$) { ...}

> function *functionname*(*arg*, *arg* , $|...$) { ...}

The vertical (or) bar separates required arguments from optional arguments in function declarations. The bar may appear at most once.

Description

Mata functions may have various numbers of arguments. How you write programs that allow these optional arguments is described below.

Remarks

Remarks are presented under the following headings:

> *What are optional arguments?*
> *How to code optional arguments*
> *Examples revisited*

What are optional arguments?

▷ Example 1

You write a function named ditty(). Function ditty() allows the caller to specify two or three arguments:

> *real matrix* ditty(*real matrix A*, *real matrix B*, *real scalar scale*)
> *real matrix* ditty(*real matrix A*, *real matrix B*)

If the caller specifies only two arguments, results are as if the caller had specified the third argument equal to missing; that is, ditty(*A*, *B*) is equivalent to ditty(*A*, *B*, .)

◁

▷ Example 2

You write function gash(). Function gash() allows the caller to specify one or two arguments:

> *real matrix* gash(*real matrix A*, *real matrix B*)
> *real matrix* gash(*real matrix A*)

If the caller specifies only one argument, results are as if J(0,0,.) were specified for the second.

◁

▷ Example 3

You write function easygoing(). Function easygoing() takes three arguments but allows the caller to specify three, two, one, or even no arguments:

> *real scalar* easygoing(*real matrix A*, *real matrix B*, *real scalar scale*)
>
> *real scalar* easygoing(*real matrix A*, *real matrix B*)
>
> *real scalar* easygoing(*real matrix A*)
>
> *real scalar* easygoing()

If *scale* is not specified, results are as if *scale* = 1 were specified. If *B* is not specified, results are as if *B* = *A* were specified. If *A* is not specified, results are as if *A* = I(2) were specified.

◁

▷ Example 4

You write function midsection(). midsection() takes three arguments, but users may specify only two—the first and last—if they wish.

> *real matrix* midsection(*real matrix A*, *real vector w*, *real matrix B*)
>
> *real matrix* midsection(*real matrix A*, *real matrix B*)

If *w* is not specified, results are as if *w* = J(1,cols(A),1) was specified.

◁

How to code optional arguments

When you code

```
function nebulous(a, b, c)
{
    . . .
}
```

you are stating that function nebulous() requires three arguments. If the caller specifies fewer or more, execution will abort.

If you code

```
function nebulous(a, b, |c)
{
    . . .
}
```

you are stating that the last argument is optional. Note the vertical or bar in front of *c*.

If you code

```
function nebulous(a, |b, c)
{
    ...
}
```

you are stating that the last two arguments are optional; the user may specify one, two, or three arguments.

If you code

```
function nebulous(|a, b, c)
{
    ...
}
```

you are stating that all arguments are optional; the user may specify zero, one, two, or three arguments.

The arguments that the user does not specify will be filled in according to the arguments' type,

if the argument type is . . .	the default value will be . . .
undeclared	J(0, 0, .)
transmorphic matrix	J(0, 0, .)
real matrix	J(0, 0, .)
complex matrix	J(0, 0, 1i)
string matrix	J(0, 0, "")
pointer matrix	J(0, 0, NULL)
transmorphic rowvector	J(1, 0, .)
real rowvector	J(1, 0, .)
complex rowvector	J(1, 0, 1i)
string rowvector	J(1, 0, "")
pointer rowvector	J(1, 0, NULL)
transmorphic colvector	J(0, 1, .)
real colvector	J(0, 1, .)
complex colvector	J(0, 1, 1i)
string colvector	J(0, 1, "")
pointer colvector	J(0, 1, NULL)
transmorphic vector	J(1, 0, .)
real vector	J(1, 0, .)
complex vector	J(1, 0, 1i)
string vector	J(1, 0, "")
pointer vector	J(1, 0, NULL)
transmorphic scalar	J(1, 1, .)
real scalar	J(1, 1, .)
complex scalar	J(1, 1, C(.))
string scalar	J(1, 1, "")
pointer scalar	J(1, 1, NULL)

Also, the function args() (see [M-5] **args( )**) will return the number of arguments that the user specified.

The vertical bar can be specified only once. That is sufficient, as we will show.

Examples revisited

▷ Example 1

In this example, real matrix function ditty(*A*, *B*, *scale*) allowed real scalar *scale* to be optional. If *scale* was not specified, results were as if *scale*=. had been specified. This can be coded

```
real matrix ditty(real matrix A, real matrix B, |real scalar scale)
{
        ...
}
```

The body of the code is written just as if *scale* were not optional because, if the caller does not specify the argument, the missing argument is automatically filled in with missing, per the table above.

◁

▷ Example 2

Real matrix function gash(*A*, *B*) allowed real matrix *B* to be optional, and if not specified, *B* = J(0,0,.) was assumed. Hence, this is coded just as example 1 was coded:

```
real matrix gash(real matrix A, |real matrix B)
{
        ...
}
```

◁

▷ Example 3

Real scalar function easygoing(*A*, *B*, *scale*) allowed all arguments to be optional. *scale* = 1 was assumed, *B* = *A*, and if necessary, *A* = I(2).

```
real scalar easygoing(|real matrix A, real matrix B,
                       real scalar scale)
{
        ...
        if (args()==2) scale = 1
        else if (args==1)  {
             B = A
             scale = 1
        }
        else if (args()==0) {
             A = B = I(2)
             scale = 1 ;
        }
        ...
}
```

◁

▷ Example 4

Real matrix function midsection(A, w, B) allowed w—its middle argument—to be omitted. If w was not specified, J(1, cols(A), 1) was assumed. Here is one solution:

```
real matrix midsection(a1, a2, |a3)
{
        if (args()==3) return(midsection_u(a1,           a2,          a3))
        else                  return(midsection_u(a1, J(1,cols(a1),1), a2))
}
real matrix midsection_u(real matrix A, real vector w, real matrix B)
{
        ...
}
```

We will never tell callers about the existence of midsection_u() even though midsection_u() is our real program.

What we did above was write midsection() to take two or three arguments, and then we called midsection_u() with the arguments in the correct position.

◁

Also See

[M-2] **intro** — Language definition

Title

[M-2] pointers — Pointers

Syntax

> pointer$\big[$ *(totype)* $\big]$ $\big[$ *orgtype* $\big]$ $\big[$ function $\big]$ ...

where *totype* is

> $\big[$ *eltype* $\big]$ $\big[$ *orgtype* $\big]$ $\big[$ function $\big]$

and where *eltype* and *orgtype* are

eltype	*orgtype*
transmorphic	matrix
numeric	vector
real	rowvector
complex	colvector
string	scalar
pointer$\big[$ *(towhat)* $\big]$	

pointer$\big[$ *(totype)* $\big]$ $\big[$ *orgtype* $\big]$ can be used in front of declarations, be they function declarations, argument declarations, or variable definitions.

Description

Pointers are objects that contain the addresses of other objects. The $*$ prefix operator obtains the contents of an address. Thus if p is a pointer, $*p$ refers to the contents of the object to which p points. Pointers are an advanced programming concept. Most programs, including involved and complicated ones, can be written without them.

In Mata, pointers are commonly used to

1. put a collection of objects under one name and

2. pass functions to functions.

One need not understand everything about pointers merely to pass functions to functions; see [M-2] **ftof**.

Remarks

Remarks are presented under the following headings:

What is a pointer?

A pointer is the address of a variable or a function. Say that variable X contains a matrix. Another variable p might contain 137,799,016, and if 137,799,016 were the address at which X were stored, then p would be said to point to X. Addresses are seldom written in base 10, so rather than saying p contains 137,799,016, we would be more likely to say that p contains 0x836a568, which is the way we write numbers in base 16. Regardless of how we write addresses, however, p contains a number and that number corresponds to the address of another variable.

In our program, if we refer to p, we are referring to p's contents, the number 0x836a568. The monadic operator * is defined as "refer to the contents of the address" or "dereference": *p means X. We could code Y = *p or Y = X, and either way, we would obtain the same result. In our program, we could refer to $X[i,j]$ or $(*p)[i,j]$, and either way, we would obtain the i, j element of X.

The monadic operator & is how we put addresses into p. To load p with the address of X, we code $p = \&X$.

The special address 0 (zero, written in hexadecimal as 0x0), also known as NULL, is how we record that a pointer variable points to nothing. A pointer variable contains NULL or it contains the address of another variable.

Or it contains the address of a function. Say that p contains 0x836a568 and that 0x836a568, rather than being the address of matrix X, is the address of function $f()$. To get the address of $f()$ into p, just as we coded $p = \&X$ previously, we code $p = \&f()$. The () at the end tells & that we want the address of a function. We code () on the end regardless of the number of arguments $f()$ requires because we are not executing $f()$, we are just obtaining its address.

To execute the function at 0x836a568—now we will assume that $f()$ takes two arguments and call them i and j—we code $(*p)(i, j)$ just as we coded $(*p)[i, j]$ when p contained the address of matrix X.

Pointers to variables

To create a pointer p to a variable, you code

   p = &*varname*

For instance, if X is a matrix,

   p = &X

stores in p the address of X. Subsequently, referring to $*p$ and referring to X amount to the same thing. That is, if X contained a 3×3 identity matrix and you coded

   $*p$ = Hilbert(4)

then after that you would find that X contained the 4×4 Hilbert matrix. X and $*p$ are the same matrix.

If X contained a 3×3 identity matrix and you coded

   ($*p$)[2,3] = 4

you would then find X[2,3] equal to 4.

You cannot, however, point to the interior of objects. That is, you cannot code

   p = &X[2,3]

and get a pointer that is equivalent to X[2,3] in the sense that if you later coded $*p$=2, you would see the change reflected in X[2,3]. The statement p = &X[2,3] is valid, but what it does, we will explain in *Pointers to expressions* below.

By the way, variables can be sustained by being pointed to. Consider the program

```
pointer(real matrix) scalar example(real scalar n)
{
    real matrix    tmp

    tmp = I(3)
    return(&tmp)
}
```

Ordinarily, variable tmp would be destroyed when example() concluded execution. Here, however, tmp's existence will be sustained because of the pointer to it. We might code

   p = example(3)

and the result will be to create $*p$ containing the 3×3 identity matrix. The memory consumed by that matrix will be freed when it is no longer being pointed to, which will occur when p itself is freed, or, before that, when the value of p is changed, perhaps by

   p = NULL

For a discussion of pointers to structures, see [M-2] **struct**.

Pointers to expressions

You can code

$$p = \&(2+3)$$

and the result will be to create $*p$ containing 5. Mata creates a temporary variable to contain the evaluation of the expression and sets p to the address of the temporary variable. That temporary variable will be freed when p is freed or, before that, when the value of p is changed, just as `tmp` was freed in the example in the previous section.

When you code

$$p = \&X[2,3]$$

the result is the same. The expression is evaluated and the result of the expression stored in a temporary variable. That is why subsequently coding $*p=2$ does not change $X[2,3]$. All $*p=2$ does is change the value of the temporary variable.

Setting pointers equal to the value of expressions can be useful. In the following code fragment, we create n 5×5 matrices for later use:

```
pvec = J(1, n, NULL)
for (i=1; i<=n; i++) pvec[i] = &(J(5, 5, .))
```

Pointers to functions

When you code

$$p = \&functionname()$$

the address of the function is stored in p. You can later execute the function by coding

$$\dots \ (*p)(\dots)$$

Distinguish carefully between

$$p = \&functionname()$$

and

$$p = \&(functionname())$$

The latter would execute *functionname*() with no arguments and then assign the returned result to a temporary variable.

For instance, assume that you wish to write a function `neat()` that will calculate the derivative of another function, which function you will pass to `neat()`. Your function, we will pretend, returns a real scalar.

You could do that as follows

```
real scalar neat(pointer(function) p, other args...)
{
        ...
}
```

although you could be more explicit as to the characteristics of the function you are passed:

```
real scalar neat(pointer(real scalar function) p, other args...)
{
    ...
}
```

In any case, inside the body of your function, where you want to call the passed function, you code

(*p) (*arguments*)

For instance, you might code

```
approx = ( (*p)(x+delta)-(*p)(x) ) / delta
```

The caller of your `neat()` function, wanting to use it with, say, function `zeta_i_just_wrote()`, would code

result = `neat(&zeta_i_just_wrote()`, *other args...*)

Pointers to pointers

Pointers to pointers (to pointers ...) are allowed, for instance, if *X* is a matrix,

p1 = &X
p2 = &p1

Here *∗p2* is equivalent to *p1*, and *∗∗p2* is equivalent to *X*.

Similarly, we can construct a pointer to a pointer to a function:

q1 = &*f*()
q2 = &p1

Here *∗q2* is equivalent to *q1*, and *∗∗q2* is equivalent to *f*().

When constructing pointers to pointers, never type &&—such as &&*x*—to obtain the address of the address of *x*. Type &(&*x*) or & &*x*. && is a synonym for &, included for those used to coding in C.

Pointer arrays

You may create an array of pointers, such as

P = (&*X1*, &*X2*, &*X3*)

or

Q = (&*f1*(), &*f2*(), &*f3*())

Here *∗P*[2] is equivalent to *X2* and *∗Q*[2] is equivalent to *f2*().

Mixed pointer arrays

You may create mixed pointer arrays, such as

$$R = (\&X, \&f())$$

Here $*R[2]$ is equivalent to $f()$.

You may not, however, create arrays of pointers mixed with real, complex, or string elements. Mata will abort with a type-mismatch error.

Definition of NULL

NULL is the special pointer value that means "points to nothing" or undefined. NULL is like 0 in many ways—for instance, coding if (X) is equivalent to coding if (X!=NULL), but NULL is distinct from zero in other ways. 0 is a numeric value; NULL is a pointer value.

Use of parentheses

Use parentheses to make your meaning clear.

In the table below, we assume

$$p = \&X$$
$$P = (\&X11, \&X12 \ \backslash \ \&X21, \&X22)$$
$$q = \&f()$$
$$Q = (\&f11(), \&f12() \ \backslash \ \&f21(), \&f22())$$

where X, $X11$, $X12$, $X21$, and $X22$ are matrices and $f()$, $f11()$, $f12()$, $f21()$, and $f22()$ are functions.

Expression	Meaning
*p	X
*p[1,1]	X
(*p)[1,1]	$X[1,1]$
*P[1,2]	$X12$
(*P[1,2])[3,4]	$X12[3,4]$
*q(a,b)	execute function $q()$ of a, b; dereference that
(*q)(a,b)	$f(a,b)$
(*q[1,1])(a,b)	$f(a,b)$
*Q[1,2](a,b)	nonsense
(*Q[1,2])(a,b)	$f12(a,b)$

Pointer arithmetic

Arithmetic with pointers (which is to say, with addresses) is not allowed:

```
: y = 2
: x = &y
: x+2
                    <stmt>:  3205  undefined operation on pointer
```

Do not confuse the expression x+2 with the expression *x+2, which is allowed and in fact evaluates to 4.

You may use the equality and inequality operators == and != with pairs of pointer values:

```
if (p1 == p2) {
    ...
}

if (p1 != p2) {
    ...
}
```

Also pointer values may be assigned and compared with the value NULL, which is much like, but still different from, zero: NULL is a 1×1 scalar containing an address value of 0. An unassigned pointer has the value NULL, and you may assign the value NULL to pointers:

```
p = NULL
```

Pointer values may be compared with NULL,

```
if (p1 == NULL) {
    ...
}

if (p1 != NULL) {
    ...
}
```

but if you attempt to dereference a NULL pointer, you will get an error:

```
: x = NULL
: *x + 2
                    <stmt>:  3120  attempt to dereference NULL pointer
```

Concerning logical expressions, you may directly examine pointer values:

```
if (p1) {
    ...
}
```

The above is interpreted as if if (p1!=NULL) were coded.

Listing pointers

You may list pointers:

```
: y = 2
: x = &y
: x
  0x8359e80
```

What is shown, 0x8359e80, is the memory address of y during our Stata session. If you typed the above lines, the address you would see could differ, but that does not matter.

Listing the value of pointers often helps in debugging because, by comparing addresses, you can determine where pointers are pointing and whether some are pointing to the same thing.

In listings, NULL is presented as 0x0.

Declaration of pointers

Declaration of pointers, as with all declarations (see [M-2] **declarations**) is optional. That basic syntax is

> pointer $\big[$ (*totype*) $\big]$ *orgtype* $\big[$ function $\big]$ ...

For instance,

> pointer(real matrix) scalar *p1*

declares that *p1* is a pointer scalar and that it points to a real matrix, and

> pointer(complex colvector) rowvector *p2*

declares that *p2* is a rowvector of pointers and that each pointer points to a complex colvector, and

> pointer(real scalar function) scalar *p3*

declares that *p3* is a pointer scalar, and that it points to a function that returns a real scalar, and

> pointer(pointer(real matrix function) rowvector) colvector *p4*

declares that *p4* is a column vector of pointers to pointers, the pointers to which each element points are rowvectors, and each of those elements points to a function returning a real matrix.

You can omit the pieces you wish.

> pointer() scalar *p5*

declares that *p5* is a pointer scalar—to what being uncertain.

> pointer scalar *p5*

means the same thing.

> pointer *p6*

declares that *p6* is a pointer, but whether it is a matrix, vector, or scalar, is unsaid.

Use of pointers to collect objects

Assume that you wish to write a function in two parts: `result_setup()` and `repeated_result()`.

In the first part, `result_setup()`, you will be passed a matrix and a function by the user, and you will make a private calculation that you will use later, each time `repeated_result()` is called. When `repeated_result()` is called, you will need to know the matrix, the function, and the value of the private calculation that you previously made.

One solution is to adopt the following design. You request the user code

```
resultinfo = result_setup(setup args...)
```

on the first call, and

```
value = repeated_result(resultinfo, other args...)
```

on subsequent calls. The design is that you will pass the information between the two functions in `resultinfo`. Here `resultinfo` will need to contain three things: the original matrix, the original function, and the private result you calculated. The user, however, never need know the details, and you will simply request that the user declare `resultinfo` as a pointer vector.

Filling in the details, you code

```
pointer vector result_setup(real matrix X, pointer(function) f)
{
        real matrix     privmat
        pointer vector  info

        . . .
        privmat = . . .
        . . .
        info = (&X, f, &privmat)
        return(info)
}

real matrix repeated_result(pointer vector info, ...)
{
        pointer(function) scalar    f
        pointer(matrix)   scalar    X
        pointer(matrix)   scalar    privmat

        f = info[2]
        X = info[1]
        privmat = info[3]

        . . .
        . . . (*f)(...) . . .
        . . . (*X) . . .
        . . . (*privmat) . . .
        . . .
}
```

It was not necessary to unload info[] into the individual scalars. The lines using the passed values could just as well have read

```
... (*info[2])(...) ...
... (*info[1]) ...
... (*info[3]) ...
```

Efficiency

When calling subroutines, it is better to pass the evaluation of pointer scalar arguments rather than the pointer scaler itself, because then the subroutine can run a little faster. Say that p points to a real matrix. It is better to code

```
... mysub(*p) ...
```

rather than

```
... mysub(p) ...
```

and then to write mysub() as

```
function mysub(real matrix X)
{
        ... X ...
}
```

rather than

```
function mysub(pointer(real matrix) scalar p)
{
        ... (*p) ...
}
```

Dereferencing a pointer (obtaining $*p$ from p) does not take long, but it does take time. Passing $*p$ rather than p can be important if mysub() loops and performs the evaluation of $*p$ hundreds of thousands or millions of times.

Diagnostics

The prefix operator * (called the dereferencing operator) aborts with error if it is applied to a nonpointer object.

Arithmetic may not be performed on undereferenced pointers. Arithmetic operators abort with error.

The prefix operator & aborts with error if it is applied to a built-in function.

Also See

[M-2] **intro** — Language definition

Title

Syntax

> pragma unset *varname*
>
> pragma unused *varname*

Description

pragma informs the compiler of your intentions so that the compiler can avoid presenting misleading warning messages and so that the compiler can better optimize the code.

Remarks

Remarks are presented under the following headings:

> *pragma unset*
> *pragma unused*

pragma unset

The pragma

> pragma unset X

suppresses the warning message

> note: variable X may be used before set

The pragma has no effect on the resulting compiled code.

In general, the warning message flags logical errors in your program, such as

```
: function problem(real matrix a, real scalar j)
> {
>     real scalar i
>
>     j = i
>     ...
> }
note: variable i may be used before set
```

Sometimes, however, the message is misleading:

```
: function notaproblem(real matrix a, real scalar j)
> {
>     real matrix V
>
>     st_view(V, ...)
>     ...
> }
note: variable V may be used before set
```

In the above, function `st_view()` (see [M-5] **st_view( )**) defines V, but the compiler does not know that.

The warning message causes no problem but, if you wish to suppress it, change the code to read

```
: function notaproblem(real matrix a, real scalar j)
> {
>       real matrix V
>
>       pragma unset V
>       st_view(V, ...)
>       ...
> }
```

`pragma unset V` states that you know V is unset and that, for warning messages, the compiler should act as if V were set at this point in your code.

pragma unused

The pragma

```
pragma unused X
```

suppresses the warning messages

```
note: argument X unused
note: variable X unused
note: variable X set but not used
```

The pragma has no effect on the resulting compiled code.

Intentionally unused variables most often arise with respect to function arguments. You code

```
: function resolve(A, B, C)
> {
>       ...
> }
note: argument C unused
```

and you know well that you are not using C. You include the unnecessary argument because you are attempting to fit into a standard or you know that, later, you may wish to change the function to include C. To suppress the warning message, change the code to read

```
: function resolve(A, B, C)
> {
>       ...
>       pragma unused C
>       ...
> }
```

The pragma states that you know C is unused and, for the purposes of warning messages, the compiler should act as if C were used at this point in your code.

Unused variables can also arise, and in general, they should simply be removed,

```
: function resin(X, Y)
> {
>     real scalar i
>     ...
>     ... code in which i never appears
>     ...
> }
note: variable i unused
```

Rather than using the pragma to suppress the message, you should remove the line `real scalar i`.

Warnings are also given for variables that are set and not used:

```
: function thwart(X, Y)
> {
>     real scalar i
>     ...
>     i = 1
>     ...
>     ... code in which i never appears
>     ...
> }
note: variable i set but unused
```

Here you should remove both the `real scalar i` and `i = 1` lines.

It is possible, however, that the set-but-unused variable was intentional:

```
: function thwart(X, Y)
> {
>     real scalar i
>     ...
>     i = somefunction(...)
>     ...
>     ... code in which i never appears
>     ...
> }
note: variable i set but not used
```

You assigned the value of `somefunction()` to `i` to prevent the result from being displayed. Here you could use `pragma unused i` to suppress the warning message, but a better alternative would be

```
: function thwart(X, Y)
> {
>     ...
>     (void) somefunction(...)
>     ...
> }
```

See *Assignment suppresses display, as does (void)* in [M-2] **exp**.

Also See

[M-2] **intro** — Language definition

Title

[M-2] **reswords** — Reserved words

Syntax

Reserved words are

aggregate	float	pointer	union
array	for	polymorphic	unsigned
	friend	pragma	using
boolean	function	private	
break		protected	vector
byte	global	public	version
	goto		virtual
case		quad	void
catch	if		volatile
class	inline	real	
colvector	int	return	while
complex		rowvector	
const	local		
continue	long	scalar	
		short	
default	mata	signed	
delegate	matrix	static	
delete		string	
do	namespace	struct	
double	new	super	
	NULL	switch	
else	numeric		
eltypedef		template	
end	operator	this	
enum	orgtypedef	throw	
explicit		transmorphic	
export		try	
external		typedef	
		typename	

Description

Reserved words are words reserved by the Mata compiler; they may not be used to name either variables or functions.

Remarks

Remarks are presented under the following headings:

Future developments
Version control

149

Future developments

Many of the words above are reserved for the future implementation of new features. For instance, the words `aggregate`, `array`, `boolean`, `byte`, etc., currently play no role in Mata, but they are reserved.

You cannot infer much about short-run development plans from the presence or absence of a word from the list. The list was constructed by including words that would be needed to add certain features, but we have erred on the side of reserving too many words because it is better to give back a word than to take it away later. Taking away a word can cause previously written code to break.

Also, features can be added without reserving words if the word will be used only within a specific context. Our original list was much longer, but then we struck from it such context-specific words.

Version control

Even if we should need to reserve new words in the future, you can ensure that you need not modify your programs by engaging in version control; see [M-2] **version**.

Also See

[M-1] **naming** — Advice on naming functions and variables

[M-2] **version** — Version control

[M-2] **intro** — Language definition

Title

[M-2] **return** — return and return(exp)

Syntax

```
return

return(exp)
```

Description

`return` causes the function to stop execution and return to the caller, returning nothing.

`return(exp)` causes the function to stop execution and return to the caller, returning the evaluation of *exp*.

Remarks

Remarks are presented under the following headings:

> *Functions that return results*
> *Functions that return nothing (void functions)*

Functions that return results

`return(exp)` specifies the value to be returned. For instance, you have written a program to return the sum of two numbers:

```
function mysum(a, b)
{
        return(a+b)
}
```

`return(exp)` may appear multiple times in the program. The following program calculates *x* factorial; it assumes *x* is an integer greater than 0:

```
real scalar myfactorial(real scalar x)
{
        if (x<=0) return(1)
        return(x*factorial(x-1))
}
```

If $x \leq 0$, the function returns 1; execution does not continue to the next line.

Functions that return a result always include one or more `return(exp)` statements.

Functions that return nothing (void functions)

A function is said to be void if it returns nothing. The following program changes the diagonal of a matrix to be 1:

```
function fixdiag(matrix A)
{
    real scalar    i

    for (i=1; i<=rows(A); i++) A[i,i] = 1
}
```

This function does not even include a return statement; execution just ends. That is fine, although the function could just as well read

```
function fixdiag(matrix A)
{
    real scalar    i

    for (i=1; i<=rows(A); i++) A[i,i] = 1
    return
}
```

The use of return is when the function has reason to end early:

```
void fixmatrix(matrix A, scalar how)
{
    real scalar    i, j

    for (i=1; i<=rows(A); i++) A[i,i] = 1
    if (how==0) return
    for (i=1; i<=rows(A); i++) {
        for (j=1; j<i; j++) A[i,j] = 0
    }
}
```

Also See

[M-5] **exit( )** — Terminate execution

[M-2] **intro** — Language definition

Title

Syntax

> *stmt*
>
> *stmt* ;

Description

Mata allows, but does not require, semicolons.

Use of semicolons is discussed below, along with advice on the possible interactions of Stata's #delimit instruction; see [P] **#delimit**.

Remarks

Remarks are presented under the following headings:

> *Optional use of semicolons*
> *You cannot break a statement anywhere even if you use semicolons*
> *Use of semicolons to create multistatement lines*
> *Significant semicolons*
> *Do not use #delimit ;*

Optional use of semicolons

You can code your program to look like this

```
real scalar foo(real matrix A)
{
    real scalar    i, sum

    sum = 0
    for (i=1; i<=rows(A); i++) {
        sum = sum + A[i,i]
    }
    return(sum)
}
```

or you can code your program to look like this:

```
real scalar foo(real matrix A)
{
    real scalar    i, sum ;

    sum = 0 ;
    for (i=1; i<=rows(A); i++) {
        sum = sum + A[i,i] ;
    }
    return(sum) ;
}
```

That is, you may omit or include semicolons at the end of statements. It makes no difference. You can even mix the two styles:

```
real scalar foo(real matrix A)
{
        real scalar    i, sum ;

        sum = 0 ;
        for (i=1; i<=rows(A); i++) {
               sum = sum + A[i,i]
        }
        return(sum)
}
```

You cannot break a statement anywhere even if you use semicolons

Most languages that use semicolons follow the rule that a statement continues up to the semicolon.

Mata follows a different rule: a statement continues across lines until it looks to be complete, and semicolons force the end of statements.

For instance, consider the statement x=b-c appearing in some program. In the code, might appear

```
x = b -
c
```

or

```
x = b -
c ;
```

and, either way, Mata will understand the statement to be x=b-c, because the statement could not possibly end at the minus: x=b- makes no sense.

On the other hand,

```
x = b
- c
```

would be interpreted by Mata as two statements: x=b and -c because x = b looks like a completed statement to Mata. The first statement will assign b to *x*, and the second statement will display the negative value of c.

Adding a semicolon will not help:

```
x = b
- c ;
```

x = b is still, by itself, a complete statement. All that has changed is that the second statement ends in a semicolon, and that does not matter.

Thus remember always to break multiline statements at places where the statement could not possibly be interpreted as being complete, such as

```
x = b -
        c + (d
                + e)
myfunction(A,
                B, C,
                    )
```

Do this whether or not you use semicolons.

Use of semicolons to create multistatement lines

Semicolons allow you to put more than one statement on a line. Rather than coding

```
a = 2
b = 3
```

you can code

```
a = 2 ;   b = 3 ;
```

and you can even omit the trailing semicolon:

```
a = 2 ;   b = 3
```

Whether you code separate statements on separate lines or the same line is just a matter of style; it does not change the meaning. Coding

```
for (i=1; i<n; i++) a[i] = -a[i] ; sum = sum + a[i] ;
```

still means

```
for (i=1; i<n; i++) a[i] = -a[i] ;
sum = sum + a[i] ;
```

and, without doubt, the programmer intended to code

```
for (i=1; i<n; i++) {
        a[i] = -a[i] ;
        sum = sum + a[i] ;
}
```

which has a different meaning.

Significant semicolons

Semicolons are not all style. The syntax for the for statement is (see [M-2] **for**)

```
for (exp1; exp2; exp3) stmt
```

Say that the complete for loop that we want to code is

```
for (x=init(); !converged(x); iterate(x))
```

and that there is no *stmt* following it. Then we must code

```
for (x=init(); !converged(x); iterate(x)) ;
```

Here we use the semicolon to force the end of the statement. Say we omitted it, and the code read

```
...
for (x=init(); !converged(x); iterate(x))
x = -x
...
```

The `for` statement would look incomplete to Mata, so it would interpret our code as if we had coded

```
for (x=init(); !converged(x); iterate(x)) {
        x = -x
}
```

Here the semicolon is significant.

Significant semicolons only happen following `for` and `while`.

Do not use #delimit ;

What follows has to do with Stata's `#delimit ;` mode. If you do not know what it is or if you never use it, you can skip what follows.

Stata has an optional ability to allow its lines to continue up to semicolons. In Stata, you code

```
. #delimit ;
```

and the delimiter is changed to semicolon until your do-file or ado-file ends, or until you code

```
. #delimit cr
```

We recommend that you do not use Mata when Stata is in `#delimit ;` mode. Mata will not mind, but you will confuse yourself.

When Mata gets control, if `#delimit ;` is on, Mata turns it off temporarily, and then Mata applies its own rules, which we have summarized above.

Also See

[M-2] **intro** — Language definition

[P] **#delimit** — Change delimiter

Title

Syntax

```
struct structname {
      declaration(s)
}
```

such as

```
struct mystruct {
      real scalar     n1, n2
      real matrix     X
}
```

Description

A structure contains a set of variables tied together under one name.

Structures may not be used interactively. They are used in programming.

Remarks

Remarks are presented under the following headings:

> *Introduction*
> *Structures and functions must have different names*
> *Structure variables must be explicitly declared*
> *Declare structure variables to be scalars whenever possible*
>
> *Vectors and matrices of structures*
> *Structures of structures*
> *Pointers to structures*
>
> *Operators and functions for use with structure members*
> *Operators and functions for use with entire structures*
>
> *Listing structures*
>
> *Use of transmorphics as passthrus*
> *Saving compiled structure definitions*
> *Saving structure variables*

Introduction

Here is an overview of the use of structures:

```
struct mystruct {
        real scalar     n1, n2
        real matrix     X
}
```

```
function myfunc()
{
        struct mystruct scalar e

        ...
        e.n1 = ...
        e.n2 = ...
        e.X  = ...
        ...
        ... mysubroutine(e, ...)
        ...
}

function mysubroutine(struct mystruct scalar x, ...)
{
        struct mystruct scalar   y
        ...
        ... x.n1 ... x.n2 ... x.X ...
        ...
        y = mysubfcn(x)
        ...
        ... y.n1 ... y.n2 ... y.X ...
        ... x.n1 ... x.n2 ... x.X ...
        ...
}

struct mystruct scalar mysubfcn(struct mystruct scalar x)
{
        struct mystruct scalar   result

        result = x
        ... result.n1 ... result.n2 ... result.X ...
        return(result)
}
```

Note the following:

1. We first defined the structure. Definition does not create variables; definition defines what we mean when we refer to a `struct mystruct` in the future. This definition is done outside and separately from the definition of the functions that will use it. The structure is defined before the functions.

2. In `myfunc()`, we declared that variable e is a `struct mystruct scalar`. We then used variables `e.n1`, `e.n2`, and `e.X` just as we would use any other variable. In the call to `mysubroutine()`, however, we passed the entire e structure.

3. In `mysubroutine()`, we declared that the first argument received is a `struct mystruct scalar`. We chose to call it x rather than e to emphasize that names are not important. y is also a `struct mystruct scalar`.

4. `mysubfcn()` not only accepts a `struct mystruct` as an argument but also returns a `struct mystruct`. One of the best uses of structures is as a way to return multiple, related values.

The line `result=x` copied all the values in the structure; we did not need to code `result.n1=x.n1`, `result.n2=x.n2`, and `result.X=x.X`.

Structures and functions must have different names

You define structures much as you define functions, at the colon prompt, with the definition enclosed in braces:

```
: struct twopart {
>         real scalar    n1, n2
> }
: function setuphistory()
> {
>         . . .
> }
```

Structures and functions may not have the same names. If you call a structure `twopart`, then you cannot have a function named `twopart()`, and vice versa.

Structure variables must be explicitly declared

Declarations are usually optional in Mata. You can code

```
real matrix swaprows(real matrix A, real scalar i1, real scalar i2)
{
        real matrix     B
        real rowvector  v

        B = A
        v = B[i1, .]
        B[i1, .] = B[i2, .]
        B[i2, .] = v
        return(B)
}
```

or you can code

```
function swaprows(A, i1, i2)
{
        B = A
        v = B[i1, .]
        B[i1, .] = B[i2, .]
        B[i2, .] = v
        return(B)
}
```

When a variable, argument, or returned value is a structure, however, you must explicitly declare it:

```
function makecalc()
{
    struct twopart scalar    t
    t.n1 = t.n2 = 0
    ...
}

function clear_twopart(struct twopart scalar t)
{
    t.n1 = t.n2 = 0
}

struct twopart scalar new_twopart()
{
    struct twopart scalar    t
    t.n1 = t.n2 = 0
    return(t)
}
```

In the functions above, we refer to variables t.n1 and t.n2. The Mata compiler cannot interpret those names unless it knows the definition of the structure.

Aside: all structure references are resolved at compile time, not run time. That is, terms like t.n1 are not stored in the compiled code and resolved during execution. Instead, the Mata compiler accesses the structure definition when it compiles your code. The compiler knows that t.n1 refers to the first element of the structure and generates efficient code to access it.

Declare structure variables to be scalars whenever possible

In our declarations, we code things like

```
struct twopart scalar    t
```

and do not simply code

```
struct twopart            t
```

although the simpler statement would be valid.

Structure variables can be scalars, vectors, or matrices; when you do not say which, matrix is assumed.

Most uses of structures are as scalars, and the compiler will generate more efficient code if you tell it that the structures are scalars. Also, when you use structure vectors or matrices, there is an extra step you need to fill in, as described in the next section.

Vectors and matrices of structures

Just as you can have real scalars, vectors, or matrices, you can have structure scalars, vectors, or matrices. The following are all valid:

```
struct twopart scalar      t
struct twopart vector      t
struct twopart rowvector   t
struct twopart colvector   t
struct twopart matrix      t
```

In a struct twopart matrix, every element of the matrix is a separate structure. Say that the matrix were 2×3. Then you could refer to any of the following variables,

```
t[1,1].n1
t[1,2].n1
t[1,3].n1
t[2,1].n1
t[2,2].n1
t[2,3].n1
```

and similarly for t$[i, j]$.n2.

If struct twopart also contained a matrix X, then

```
t[i, j].X
```

would refer to the (i, j)th matrix.

```
t[i, j].X[k, l]
```

would refer to the (k, l)th element of the (i, j)th matrix.

If t is to be a 2×3 matrix, you must arrange to make it 2×3. After the declaration

```
struct twopart matrix    t
```

t is 0×0. This result is no different from the situation where t is a real matrix and after declaration, t is 0×0.

Whether t is a real matrix or a struct twopart matrix, you allocate t by assignment. Let's pretend that t is a real matrix. There are three solutions to the allocation problem:

(1) t = x

(2) t = somefunction(...)

(3) t = J(r, c, v)

All three are so natural that you do not even think of them as allocation; you think of them as definition.

The situation with structures is the same.

Let's take each in turn.

1. x contains a 2×3 struct twopart. You code

```
t = x
```

and now t contains a copy of x. t is 2×3.

2. somefunction(...) returns a 2 × 3 struct twopart. You code

> t = somefunction(...)

and now t contains the 2 × 3 result.

3. Mata function J(r, c, v) returns an $r \times c$ matrix, every element of which is set to v. So pretend that variable tpc contains a struct twopart scalar. You code

> t = J(2, 3, tpc)

and now t is 2 × 3, every element of which is a copy of tpc. Here is how you might do that:

```
function ...(...)
{
        struct twopart scalar    tpc
        struct twopart matrix    t

        ...
        t = J(2, 3, tpc)
        ...
}
```

Finally, there is a fourth way to create structure vectors and matrices. When you defined

```
struct twopart {
        real scalar     n1, n2
}
```

Mata not only recorded the structure definition but also created a function named twopart() that returns a struct twopart. Thus, rather than enduring all the rigamarole of creating a matrix from a preallocated scalar, you could simply code

> t = J(2, 3, twopart())

In fact, the function twopart() that Mata creates for you allows 0, 1, or 2 arguments:

twopart()   returns a 1 × 1 struct twopart

twopart(r)   returns an $r \times 1$ struct twopart

twopart(r, c)   returns an $r \times c$ struct twopart

so you could code

> t = twopart(2, 3)

or you could code

> t = J(2, 3, twopart())

and whichever you code makes no difference.

Either way, what is in t? Each element contains a separate struct twopart. In each struct twopart, the scalars have been set to missing (., "", or NULL, as appropriate), the vectors and row vectors have been made 1 × 0, the column vectors 0 × 1, and the matrices 0 × 0.

Structures of structures

Structures may contain other structures:

```
struct twopart {
        real scalar     n1, n2
}

struct pair_of_twoparts {
        struct twopart scalar   t1, t2
}
```

If t were a struct pair_of_twoparts scalar, then the members of t would be

```
t.t1        a struct twopart scalar
t.t2        a struct twopart scalar
t.t1.n1     a real scalar
t.t1.n2     a real scalar
t.t2.n1     a real scalar
t.t2.n2     a real scalar
```

You may also create structures of structures of structures, structures of structures of structures of structures, and so on. You may not, however, include a structure in itself:

```
struct recursive {
        ...
        struct recursive scalar   r
        ...
}
```

Do you see how, even in the scalar case, struct recursive would require an infinite amount of memory to store?

Pointers to structures

What you can do is this:

```
struct recursive {
        ...
        pointer(struct recursive scalar) scalar   r
        ...
}
```

Thus, if r were a struct recursive scalar, then *r.r would be the next structure, or r.r would be NULL if there were no next structure. Immediately after allocation, r.r would equal NULL.

In this way, you may create linked lists.

Mata provides operator -> for accessing members of pointers to structures.

Let rec be a struct recursive, and assume that struct recursive also had member real scalar n, so that rec.n would be rec's n value. The value of the next structure's n would be rec.r->n (assuming rec.r!=NULL).

The syntax of `->` is

> *exp1* `->` *exp2*

where *exp1* evaluates to a structure pointer and *exp2* indexes the structure.

Operators and functions for use with structure members

All operators, all functions, and all features of Mata work with members of structures. That is, given

```
struct example {
        real scalar n
        real matrix X
}

function ... (...)
{
        real scalar rs
        real matrix rm
        struct example scalar ex

        ...
}
```

then `ex.n` and `ex.X` may be used anyplace `rs` and `rm` would be valid.

Operators and functions for use with entire structures

Some operators and functions can be used with entire structures, not just the structure's elements. Given

```
struct mystruct scalar    ex1, ex2, ex3, ex4
struct mystruct matrix    E, F, G
```

1. You may use `==` and `!=` to test for equality:

> `if (ex1==ex2) ...`

> `if (ex1!=ex2) ...`

Two structures are equal if their members are equal.

In the example, `struct mystruct` itself contains no substructures. If it did, the definition of equality would include checking the equality of substructures, sub-substructures, etc.

In the example, `ex1` and `ex2` are scalars. If they were matrices, each element would be compared, and the matrices would be equal if the corresponding elements were equal.

2. You may use `:==` and `:!=` to form pattern matrices of equality and inequality.

3. You may use the comma and backslash operators to form vectors and matrices of structures:

> `ex = ex1, ex2 \ ex3, ex4`

4. You may use `&` to obtain pointers to structures:

> `ptr_to_ex1 = &ex1`

5. You may use subscripting to access and copy structure members:

```
ex1 = E[1,2]
E[1,2] = ex1
F = E[2,.]
E[2,.] = F
G = E[|1,1\2,2|]
E[|1,1\2,2|] = G
```

6. You may use the `rows()` and `cols()` functions to obtain the number of rows and columns of a matrix of structures.

7. You may use `eltype()` and `orgtype()` with structures. `eltype()` returns `struct`; `orgtype()` returns the usual results.

8. You may use most functions that start with the letters *is*, as in `isreal()`, `iscomplex()`, `isstring()`, etc. These functions return 1 if true and 0 if false and with structures, usually return 0.

9. You may use `swap()` with structures.

Listing structures

To list the contents of a structure variable, as for debugging purposes, use function `liststruct()`; see [M-5] **liststruct( )**.

Using the default, unassigned-expression method to list structures is not recommended, because all that is shown is a pointer value instead of the structure itself.

Use of transmorphics as passthrus

A `transmorphic matrix` can theoretically hold anything, so when we told you that structures had to be explicitly declared, that was not exactly right. Say that function `twopart()` returns a `struct twopart` scalar. You could code

```
x = twopart()
```

without declaring x (or declaring it `transmorphic`), and that would not be an error. What you could not do would be to then refer to x.n1 or x.n2, because the compiler would not know that x contains a `struct twopart` and so would have no way of interpreting the variable references.

This property can be put to good use in implementing handles and passthrus.

Say that you are implementing a complicated system. Just to fix ideas, we'll pretend that the system finds the maximum of user-specified functions and that the system has many bells and whistles. To track a given problem, let's assume that your system needs many variables. One variable records the method to be used. Another records whether numerical derivatives are to be used. Another records the current gradient vector. Another records the iteration count, and so on. There might be hundreds of these variables.

You bind all these variables together in one structure:

```
struct maxvariables {
        real scalar   method
        real scalar   use_numeric_d
        real vector   gradient
        real scalar   iteration
        ...
}
```

You design a system with many functions, and some functions call others, but since all the status variables are bound together in one structure, it is easy to pass the values from one function to another.

You also design a system that is easy to use. It starts by the user "opening" a problem,

```
handle = maximize_open()
```

and from that point on the user passes the handle around to the other maximize routines:

```
maximize_set_use_numeric_d(handle, 1)
maximize_set_function_to_max(handle, &myfunc())
...
maximize_maximize_my_function(handle)
```

In this way, you, the programmer of this system, can hold on to values from one call to the next, and you can change the values, too.

What you never do, however, is tell the user that the handle is a struct maxvariables. You just tell the user to open a problem by typing

```
handle = maximize_open()
```

and then to pass the handle returned to the other maximize_*() routines. If the user feels that he must explicitly declare the handle, you tell him to declare it:

```
transmorphic scalar handle
```

What is the advantage of this secrecy? You can be certain that the user never changes any of the values in your struct maxvariables because the compiler does not even know what they are.

Thus you have made your system more robust to user errors.

Saving compiled structure definitions

You save compiled structure definitions just as you save compiled function definitions; see [M-3] **mata mosave** and [M-3] **mata mlib**.

When you define a structure, such as twopart,

```
struct twopart {
        real scalar    n1, n2
}
```

that also creates a function, twopart(), that creates instances of the structure.

Saving twopart() in a .mo file, or in a .mlib library, saves the compiled definition as well. Once twopart() has been saved, you may write future programs without bothering to define struct twopart. The definition will be automatically found.

Saving structure variables

Variables containing structures may be saved on disk just as you would save any other variable. No special action is required. See [M-3] **mata matsave** and see the function `fputmatrix()` in [M-5] **fopen( )**. `mata matsave` and `fputmatrix()` both work with structure variables, although their entries do not mention them.

Also See

[M-2] **declarations** — Declarations and types

[M-2] **intro** — Language definition

Title

Syntax

$x[real\ vector\ r,\ real\ vector\ c]$

$x[|\ real\ matrix\ sub|]$

Subscripts may be used on the left or right of the equal-assignment operator.

Description

Subscripts come in two styles.

In [*subscript*] syntax—called list subscripts—an element or a matrix is specified:

x[1,2]   the 1,2 element of x; a scalar

x[(1\3\2), (4,5)]   the 3×2 matrix composed of rows 1, 3, and 2
                        and columns 4 and 5 of x:

$$\begin{bmatrix} x_{14} & x_{15} \\ x_{34} & x_{35} \\ x_{24} & x_{25} \end{bmatrix}$$

In [|*subscript*|] syntax—called range subscripts—an element or a contiguous submatrix is specified

x[|1,2|]   same as x[1,2]; a scalar

x[|2,3 \ 4,7|] 3×4 submatrix of x:

$$\begin{bmatrix} x_{23} & x_{24} & x_{25} & x_{26} & x_{27} \\ x_{33} & x_{34} & x_{35} & x_{36} & x_{37} \\ x_{43} & x_{44} & x_{45} & x_{46} & x_{47} \end{bmatrix}$$

Both style subscripts may be used in expressions and may be used on the left-hand side of the equal-assignment operator.

Remarks

Remarks are presented under the following headings:

> List subscripts
> Range subscripts
> When to use list subscripts and when to use range subscripts
> A fine distinction

List subscripts

List subscripts—also known simply as subscripts—are obtained when you enclose the subscripts in square brackets, [and]. List subscripts come in two basic forms:

$x[ivec, jvec]$	matrix composed of rows *ivec* and columns *jvec* of matrix x
$v[kvec]$	vector composed of elements *kvec* of vector v

where *ivec*, *jvec*, *kvec* may be a vector or a scalar, so the two basic forms include

$x[i, j]$	scalar i, j element
$x[i, jvec]$	row vector of row i, elements *jvec*
$x[ivec, j]$	column vector of column j, elements *ivec*
$v[k]$	scalar kth element of vector v

Also missing value may be specified to mean all the rows or all the columns:

$x[i, .]$	row vector of row i of x
$x[., j]$	column vector of column j of x
$x[ivec, .]$	matrix of rows *ivec*, all columns
$x[., jvec]$	matrix of columns *jvec*, all rows
$x[., .]$	the entire matrix

Finally, Mata assumes missing value when you omit the argument entirely:

$x[i,]$	same as $x[i, .]$
$x[ivec,]$	same as $x[ivec, .]$
$x[, j]$	same as $x[., j]$
$x[, jvec]$	same as $x[., jvec]$
$x[,]$	same as $x[., .]$

Good style is to specify *ivec* as a column vector and *jvec* as a row vector, but that is not required:

$x[(1\backslash2\backslash3), (1,2,3)]$	good style
$x[(1,2,3), (1,2,3)]$	same as $x[(1\backslash2\backslash3), (1,2,3)]$
$x[(1\backslash2\backslash3), (1\backslash2\backslash3)]$	same as $x[(1\backslash2\backslash3), (1,2,3)]$
$x[(1,2,3), (1\backslash2\backslash3)]$	same as $x[(1\backslash2\backslash3), (1,2,3)]$

Similarly, good style is to specify *kvec* as a column when v is a column vector and to specify *kvec* as a row when v is a row vector, but that is not required and what is returned is a column vector if v is a column and a row vector if v is a row:

rowv$[(1,2,3)]$	good style for specifying row vector
rowv$[(1\backslash2\backslash3)]$	same as *rowv*$[(1,2,3)]$
colv$[(1\backslash2\backslash3)]$	good style for specifying column vector
colv$[(1,2,3)]$	same as *colv*$[(1\backslash2\backslash3)]$

Subscripts may be used in expressions following a variable name:

```
first = list[1]
multiplier = x[3,4]
result = colsum(x[,j])
```

Subscripts may be used following an expression to extract a submatrix from a result:

```
allneeded = invsym(x)[(1::4), .] * multiplier
```

Subscripts may be used on the left-hand side of the equal-assignment operator:

```
x[1,1] = 1
x[1,.] = y[3,.]
x[(1::4), (1..4)] = I(4)
```

Range subscripts

Range subscripts appear inside the difficult to type [| and |] brackets. Range subscripts come in four basic forms:

$x[\|i,j\|]$	i,j element; same result as $x[i,j]$
$v[\|k\|]$	kth element of vector; same result as $v[k]$
$x[\|i,j \setminus k,l\|]$	submatrix, vector, or scalar formed using (i,j) as top-left corner and (k,l) as bottom-right corner
$v[\|i \setminus k\|]$	subvector or scalar of elements i through k; result is row vector if v is row vector, column vector if v is column vector

Missing value may be specified for a row or column to mean all rows or all columns when a 1×2 or 1×1 subscript is specified:

$x[\|i,.\|]$	row i of x; same as $x[i,.]$
$x[\|.,j\|]$	column j of x; same as $x[.,j]$
$x[\|.,.\|]$	entire matrix; same as $x[.,.]$
$v[\|.\|]$	entire vector; same as $v[.]$

Also missing may be specified to mean the number of rows or the number of columns of the matrix being subscripted when a 2×2 subscript is specified:

$x[\|1,2 \setminus 4,.\|]$	equivalent to $x[\|1,2 \setminus 4,\text{cols}(x)\|]$
$x[\|1,2 \setminus .,3\|]$	equivalent to $x[\|1,2 \setminus \text{rows}(x),3\|]$
$x[\|1,2 \setminus .,.\|]$	equivalent to $x[\|1,2 \setminus \text{rows}(x),\text{cols}(x\|]$

With range subscripts, what appears inside the square brackets is in all cases interpreted as a matrix expression, so in

```
sub = (1,2)
... x[|sub|] ...
```

`x[sub]` refers to `x[1,2]`.

Range subscripts may be used in all the same contexts as list subscripts; they may be used in expressions following a variable name

        `submat = result[|1,1 \ 3,3|]`

they may be used to extract a submatrix from a calculated result

        `allneeded = invsym(x)[|1,1 \ 4,4|]`

and they may be used on the left-hand side of the equal-assignment operator:

        `x[|1,4 \ 1,4|] = I(4)`

When to use list subscripts and when to use range subscripts

Everything a range subscript can do, a list subscript can also do. The one seemingly unique feature of a range subscript,

        $x[|i1,j1 \ i2,j2|]$

is perfectly mimicked by

        $x[(i1::i2), (j1..j2)]$

The range-subscript construction, however, executes more quickly, and so that is the purpose of range subscripts: to provide a fast way to extract contiguous submatrices. In all other cases, use list subscripts because they are faster.

Use list subscripts to refer to scalar values:

        `result = x[1,3]`
        `x[1,3] = 2`

Use list subscripts to extract entire rows or columns:

        `obs = x[., 3]`
        `var = x[4, .]`

Use list subscripts to permute the rows and columns of matrices:

        `: x = (1,2,3,4 \ 5,6,7,8 \ 9,10,11,12)`
        `: y = x[(1\3\2), .]`
        `: y`

```
        1    2    3    4

  1     1    2    3    4
  2     9   10   11   12
  3     5    6    7    8
```

        `: y = x[., (1,3,2,4)]`
        `: y`

```
        1    2    3    4

  1     1    3    2    4
  2     5    7    6    8
  3     9   11   10   12
```

        `: y=x[(1\3\2), (1,3,2,4)]`

```
: y
        1    2    3    4
   1 |  1    3    2    4  |
   2 |  9   11   10   12  |
   3 |  5    7    6    8  |
```

Use list subscripts to duplicate rows or columns:

```
: x = (1,2,3,4 \ 5,6,7,8 \ 9,10,11,12)
: y = x[(1\2\3\1), .]
: y
        1    2    3    4
   1 |  1    2    3    4  |
   2 |  5    6    7    8  |
   3 |  9   10   11   12  |
   4 |  1    2    3    4  |

: y = x[., (1,2,3,4,2)]
: y
        1    2    3    4    5
   1 |  1    2    3    4    2  |
   2 |  5    6    7    8    6  |
   3 |  9   10   11   12   10  |

: y = x[(1\2\3\1), (1,2,3,4,2)]
: y
        1    2    3    4    5
   1 |  1    2    3    4    2  |
   2 |  5    6    7    8    6  |
   3 |  9   10   11   12   10  |
   4 |  1    2    3    4    2  |
```

A fine distinction

There is a fine distinction between $x[i,j]$ and $x[|i,j|]$. In $x[i,j]$, there are two arguments, i and j. The comma separates the arguments. In $x[|i,j|]$, there is one argument: i,j. The comma is the column-join operator.

In Mata, comma means mostly the column-join operator:

```
newvec = oldvec, addedvalues
qsum = (x,1)'(x,1)
```

There are, in fact, only two exceptions. When you type the arguments for a function, the comma separates one argument from the next:

```
result = f(a,b,c)
```

In the above example, $f()$ receives three arguments: a, b, and c. If we wanted $f()$ to receive one argument, (a,b,c), we would have to enclose the calculation in parentheses:

```
result = f((a,b,c))
```

That is the first exception. When you type the arguments inside a function, comma means argument separation. You get back to the usual meaning of comma—the column-join operator—by opening another set of parentheses.

The second exception is in list subscripting:

$x[i,j]$

Inside the list-subscript brackets, comma means argument separation. That is why you have seen us type vectors inside parentheses:

$x[(1\backslash2\backslash3),(1,2,3)]$

These are the two exceptions. Range subscripting is not an exception. Thus in

$x[|\ i,\ j|]$

there is one argument, i,j. With range subscripts, you may program constructs such as

```
IJ    = (i,j)
RANGE = (1,2 \ 4,4)
...
... x[|IJ|] ... x[|RANGE|] ...
```

You may not code in this way with list subscripts. In particular, $x[IJ]$ would be interpreted as a request to extract elements i and j from vector x, and would be an error otherwise. $x[RANGE]$ would always be an error.

We said earlier that list subscripts $x[i,j]$ are a little faster than range subscripts $x[|i,j|]$. That is true, but if $IJ=(i,j)$ already, $x[|IJ|]$ is faster than $x[i,j]$. You would, however, have to execute many millions of references to $x[|IJ|]$ before you could measure the difference.

Conformability

$x[i,j]$:

x:	$r \times c$			
i:	$m \times 1$	or	$1 \times m$	(does not matter which)
j:	$1 \times n$	or	$n \times 1$	(does not matter which)
result:	$m \times n$			

$x[i, .]$:

x:	$r \times c$			
i:	$m \times 1$	or	$1 \times m$	(does not matter which)
result:	$m \times c$			

$x[., j]$:

x:	$r \times c$			
j:	$1 \times n$	or	$n \times 1$	(does not matter which)
result:	$r \times n$			

$x[., .]$:

x:	$r \times c$
result:	$r \times c$

$x[i]$:

x:	$n \times 1$		$1 \times n$
i:	$m \times 1$ or $1 \times m$		$1 \times m$ or $m \times 1$
result:	$m \times 1$		$1 \times m$

$x[.]$:

x:	$n \times 1$	$1 \times n$
result:	$n \times 1$	$1 \times n$

$x[|k|]$:

x:	$r \times c$
k:	1×2
result:	1×1 if $k[1]<.$ and $k[2]<.$
	$r \times 1$ if $k[1]>=.$ and $k[2]<.$
	$1 \times c$ if $k[1]<.$ and $k[2]>=.$
	$r \times c$ if $k[1]>=.$ and $k[2]>=.$

$x[|k|]$:

x:	$r \times c$
k:	2×2
result:	$k[2,1]-k[1,1]+1 \times k[2,2]-k[1,2]+1$
	(in the above formula, if $k[2,1]>=.$, treat as if $k[2,1]=r$,
	and similarly, if $k[2,2]>=.$, treat as if $k[2,2]=c$)

$x[|k|]$:

x:	$r \times 1$	$1 \times c$
k:	2×1	2×1
result:	$k[2]-k[1]+1 \times 1$	$1 \times k[2]-k[1]+1$
	(if $k[2]>=.$, treat as	(if $k[2]>=.$, treat as
	if $k[2]=r$)	if $k[2]=c$)

Diagnostics

Both styles of subscripts abort with error if the subscript is out of range, if a reference is made to a nonexisting row or column.

Reference

Gould, W. 2007. Mata Matters: Subscripting. *Stata Journal* 7: 106–116.

Also See

[M-2] **intro** — Language definition

Title

Syntax

The basic language syntax is

istmt

where

istmt :=	*stmt*	
	function *name* *(farglist) fstmt*	
	ftype *name(farglist) fstmt*	
	ftype function *name* *(farglist) fstmt*	
stmt :=	*nothing*	
	;	(meaning *nothing*)
	version *number*	
	{ *stmt* ... }	
	exp	
	pragma *pstmt*	
	if *(exp) stmt*	
	if *(exp) stmt* else *stmt*	
	for *(exp;exp;exp) stmt*	
	while *(exp) stmt*	
	do *stmt* while *(exp)*	
	break	
	continue	
	label:	
	goto *label*	
	return	
	return*(exp)*	
fstmt :=	*stmt*	
	type arglist	
	external *type arglist*	
arglist :=	*name*	
	name()	
	name, *arglist*	
	name(), *arglist*	
farglist :=	*nothing*	
	efarglist	
efarglist :=	*felement*	
	felement, *efarglist*	
	| *felement*	
	| *felement*, *efarglist*	

felement :=	*name*	
	type name	
	name()	
	type name()	
ftype :=	*type*	
	`void`	
type :=	*eltype*	
	orgtype	
	eltype orgtype	
eltype :=	`transmorphic`	
	`string`	
	`numeric`	
	`real`	
	`complex`	
	`pointer`	
	`pointer`(*ptrtype*)	
orgtype :=	`matrix`	
	`vector`	
	`rowvector`	
	`colvector`	
	`scalar`	
ptrtype :=	*nothing*	
	type	
	type `function`	
	`function`	
pstmt :=	`unset` *name*	
	`unused` *name*	
name :=	identifier up to 32 characters long	
label :=	identifier up to 8 characters long	
exp :=	expression as defined in [M-2] **exp**	

Description

Mata is a C-like compiled-into-pseudocode language with matrix extensions and run-time linking.

Remarks

Remarks are presented under the following headings:

        Treatment of semicolons
        Types and declarations
        Void matrices
        Void functions
        Operators
        Subscripts
        Implied input tokens
        Function argument-passing convention
        Passing functions to functions
        Optional arguments

After reading [M-2] **syntax**, see [M-2] **intro** for a list of entries that give more explanation of what is discussed here.

Treatment of semicolons

Semicolon (;) is treated as a line separator. It is not required, but it may be used to place two statements on the same physical line:

```
x = 1 ; y = 2 ;
```

The last semicolon in the above example is unnecessary but allowed.

Single statements may continue onto more than one line if the continuation is obvious. Take "obvious" to mean that there is a hanging open parenthesis or a hanging dyadic operator; e.g.,

```
x = (
        3)
x = x +
        2
```

See [M-2] **semicolons** for more information.

Types and declarations

The *type* of a variable or function is described by

> *eltype   orgtype*

where *eltype* and *orgtype* are each one of

eltype	*orgtype*
transmorphic	matrix
numeric	vector
real	rowvector
complex	colvector
string	scalar
pointer	

For example, a variable might be real scalar, or complex matrix, or string vector.

Mata also has structures—the *eltype* is struct *name*—but these are not discussed here. For a discussion of structures, see [M-2] **struct**.

Declarations are optional. When the *type* of a variable or function is not declared, it is assumed to be a transmorphic matrix. In particular:

1. *eltype* specifies the type of the elements. When *eltype* is not specified, transmorphic is assumed.

2. *orgtype* specifies the organization of the elements. When *orgtype* is not specified, matrix is assumed.

All *types* are special cases of transmorphic matrix.

The nesting of *eltypes* is

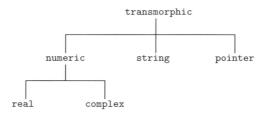

orgtypes amount to nothing more than a constraint on the number of rows and columns of a matrix:

orgtype	constraint
matrix	$r \geq 0$ and $c \geq 0$
vector	$r = 1$ and $c \geq 0$ or $r \geq 0$ and $c = 1$
rowvector	$r = 1$ and $c \geq 0$
colvector	$r \geq 0$ and $c = 1$
scalar	$r = 1$ and $c = 1$

See [M-2] **declarations**.

Void matrices

A matrix (vector, row vector, or column vector) that is 0×0, $r \times 0$, or $0 \times c$ is said to be void; see [M-2] **void**.

The function $J(r, c, val)$ returns an $r \times c$ matrix with each element containing *val*; see [M-5] **J( )**.

J( ) can be used to create void matrices.

See [M-2] **void**.

Void functions

Rather than *eltype orgtype*, a function can be declared to return nothing by being declared to return void:

```
void function example(matrix A)
{
        real scalar i

        for (i=1; i<=rows(A); i++) A[i,i] = 1
}
```

A function that returns nothing (does not include a return(*exp*) statement), in fact returns J(0, 0, .), and the above function could equally well be coded as

```
void function example(matrix A)
{
    real scalar   i
    for (i=1; i<=rows(A); i++) A[i,i] = 1
    return(J(0, 0, .))
}
```

or

```
void function example(matrix A)
{
    real scalar   i
    for (i=1; i<=rows(A); i++) A[i,i] = 1
    return(J(0,0,.))
}
```

Therefore, void also is a special case of transmorphic matrix (it is in fact a 0×0 real matrix). Since declarations are optional (but recommended both for reasons of style and for reasons of efficiency), the above function could also be coded as

```
function example(A)
{
    for (i=1; i<=rows(A); i++) A[i,i] = 1
}
```

See [M-2] **declarations**.

Operators

Mata provides the usual assortment of operators; see [M-2] **exp**.

The monadic prefix operators are

```
-   !   ++   --   &   *
```

Prefix operators & and * have to do with pointers; see [M-2] **pointers**.

The monadic postfix operators are

```
'   ++   --
```

Note the inclusion of postfix operator ' for transposition. Also, for Z complex, Z' returns the conjugate transpose. If you want the transposition without conjugation, see [M-5] **transposeonly( )**.

The dyadic operators are

```
=   ?   \   ::   ,   ..   |   &   ==   >=   <=   <   >
!=  +   -   *   #   ^
```

Also, && and || are included as synonyms for & and |.

The operators == and != do not require conformability, nor do they require that the matrices be of the same type. In such cases, the matrices are unequal (== is false and != is true). For complex arguments, <, <=, >, and >= refer to length of the complex vector. == and !=, however, refer not to length but to actual components. See [M-2] **op_logical**.

The operators , and \ are the row-join and column-join operators. (1,2,3) constructs the row vector (1,2,3). (1\2\3) constructs the column vector (1,2,3)'. (1,2\3,4) constructs the matrix with first row (1,2) and second row (3,4). a,b joins two scalars, vectors, or matrices rowwise. $a\backslash b$ joins two scalars, vectors, or matrices columnwise. See [M-2] **op_join**.

.. and :: refer to the row-to and column-to operators. 1..5 is (1,2,3,4,5). 1::5 is (1\2\3\4\5). 5..1 is (5,4,3,2,1). 5::1 is (5\4\3\2\1). See [M-2] **op_range**.

For |, &, ==, >=, <=, <, >, !=, +, −, *, /, and ^, there is :op at precedence just below op. These operators perform the elementwise operation. For instance, $A*B$ refers to matrix multiplication; $A:*B$ refers to elementwise multiplication. Moreover, elementwise is generalized to cases where A and B do not have the same number of rows and the same number of columns. For instance, if A is a $1 \times c$ row vector and B is a $r \times c$ matrix, then $||C_{ij}|| = ||A_j|| * ||B_{ij}||$ is returned. See [M-2] **op_colon**.

Subscripts

$A[i,j]$ returns the i,j element of A.

$A[k]$ returns $A[1,k]$ if A is $1 \times c$ and $A[k,1]$ if A is $r \times 1$. That is, in addition to declared vectors, any $1 \times c$ matrix or $r \times 1$ matrix may be subscripted by one index. Similarly, any vector can be subscripted by two indices.

i, j, and k may be vectors as well as scalars. For instance, $A[(3\backslash5),4]$ returns a 3×1 column vector containing rows 3 to 5 of the 4th column.

i, j, and k may be missing value. $A[.,4]$ returns a column vector of the 4th column of A.

The above subscripts are called list-style subscripts. Mata provides a second format called range-style subscripts that is especially useful for selecting submatrices. $A[|3,3\backslash5,5|]$ returns the 2×2 submatrix of A starting at $A[3,3]$.

See [M-2] **subscripts**.

Implied input tokens

Before interpreting and compiling a line, Mata makes the following substitutions to what it sees:

Input sequence	Interpretation
'*name*	'**name*
[,	[.,
,]	,.]

Hence, coding X′Z is equivalent to coding X′*Z, and coding x = z[1,] is equivalent to coding x = z[1,.].

Function argument-passing convention

Arguments are passed to functions by address, also known as by name or by reference. They are not passed by value. When you code

        ... f(A) ...

it is the address of A that is passed to f(), not a copy of the values in A. f() can modify A.

Most functions do not modify their arguments, but some do. lud(A, L, U, p), for instance, calculates the LU decomposition of A. The function replaces the contents of L, U, and p with matrices such that $L[p,]*U = A$.

Oldtimers will have heard of the FORTRAN programmer who called a subroutine and passed to it a second argument of 1. Unbeknownst to him, the subroutine changed its second argument, with the result that the constant 1 was changed throughout the rest of his code. That cannot happen in Mata. When an expression is passed as an argument (and constants are expressions), a temporary variable containing the evaluation is passed to the function. Modifications to the temporary variable are irrelevant because the temporary variable is discarded once the function returns. Thus if f() modifies its second argument and you call it by coding f(A,2), because 2 is copied to a temporary variable, the value of the literal 2 will remain unchanged on the next call.

If you call a function with an expression that includes the assignment operator, it is the left-hand side of the expression that is passed. That is, coding

        f(a, b=c)

has the same result as coding

        b = c
        f(a, b)

If function f() changes its second argument, it will be b and not c that is modified.

Also, Mata attempts not to create unnecessary copies of matrices. For instance, consider

        function changearg(x) x[1,1] = 1

changearg(mymat) changes the 1,1 element of mymat to 1. Now let us define

        function cp(x) return(x)

Coding changearg(cp(mymat)) would still change mymat because cp() returned x itself. On the other hand, if we defined cp() as

```
function cp(x)
{
      matrix t

      t = x
      return(t)
}
```

then coding changearg(cp(mymat)) would not change mymat. It would change a temporary matrix which would be discarded once changearg() returned.

Passing functions to functions

One function may receive another function as an argument using pointers. One codes

```
function myfunc(pointer(function) f, a, b)
{
        ... (*f)(a) ... (*f)(b) ...
}
```

although the `pointer(function)` declaration, like all declarations, is optional. To call `myfunc()` and tell it to use function `prima()` for `f()`, and 2 and 3 for `a` and `b`, one codes

```
myfunc(&prima(), 2, 3)
```

See [M-2] **ftof** and [M-2] **pointers**.

Optional arguments

Functions may be coded to allow receiving a variable number of arguments. This is done by placing a vertical or bar (|) in front of the first argument that is optional. For instance,

```
function mynorm(matrix A, |scalar power)
{
        ...
}
```

The above function may be called with one matrix or with a matrix followed by a scalar.

The function `args()` (see [M-5] **args( )**) can be used to determine the number of arguments received and to set defaults:

```
function mynorm(matrix A, |scalar power)
{
        ...
        (args()==1) power = 2
        ...
}
```

See [M-2] **optargs**.

Reference

Gould, W. 2005. Mata Matters: Translating Fortran. *Stata Journal* 5: 421–441.

Also See

[M-2] **intro** — Language definition

Title

Syntax

Syntax 1

```
. version #[ .# ]
. mata:
: ...
: function name(...)
: {
:          ...
: }
: ...
: end
```

Syntax 2

```
: function name(...)
: {
:          version #[ .# ]
:          ...
: }
```

Description

In syntax 1, Stata's version command (see [P] **version**) sets the version before entering Mata. This specifies both the compiler and library versions to be used. Syntax 1 is recommended.

In syntax 2, Mata's version command sets the version of the library functions that are to be used. Syntax 2 is rarely used.

Remarks

Remarks are presented under the following headings:

> *Purpose of version control*
> *Recommendations for do-files*
> *Recommendations for ado-files*
> *Compile-time and run-time versioning*

Purpose of version control

Mata is under continual development, which means not only that new features are being added but also that old features sometimes change how they work. Old features changing how they work is supposedly an improvement—it generally is—but that also means old programs might stop working or, worse, work differently.

version provides the solution.

If you are working interactively, nothing said here matters.

If you use Mata in do-files or ado-files, we recommend that you set version before entering Mata.

Recommendations for do-files

The recommendation for do-files that use Mata is the same as for do-files that do not use Mata: specify the version number of the Stata you are using on the top line:

```
─────────────────────────────────────────── top of myfile.do ───────────
      version 10
      ...
─────────────────────────────────────────── end of myfile.do ───────────
```

To determine the number that should appear after version, type about at the Stata prompt:

```
. about
Stata/SE 10.0
  (output omitted )
```

We are using Stata 10.0.

Coding version 10 will not benefit us today but, in the future, we will be able to rerun our do-file and obtain the same results.

By the way, a do-file is any file that you intend to execute using Stata's do or run commands, regardless of the file suffix. Many users (us included) save Mata source code in .mata files and then type do *myfile*.mata to compile. .mata files are do-files; we include the version line:

```
─────────────────────────────────────────── top of  myfile.mata ───────────
      version 10
      mata:
      ...
      end
─────────────────────────────────────────── end of myfile.mata ───────────
```

Recommendations for ado-files

Mata functions may be included in ado-files; see [M-1] **ado**. In such files, set version before entering Mata along with, as usual, setting the version at the top of your program:

```
─────────────────────────────────────────── top of myfile.ado ───────────
      program myfile
              version 10        ← as usual
              ...
      end

      version 10                ← new
      mata:
      ...
      end
─────────────────────────────────────────── end of myfile.ado ───────────
```

Compile-time and run-time versioning

What follows is detail. We recommend always following the recommendations above.

There are actually two version numbers that matter—the version number set at the time of compilation, which affects how the source code is interpreted, and the version of the libraries used to supply subroutines at the time of execution.

The version command that we used in the previous sections is in fact Stata's version command (see [P] **version**), and it sets both versions:

```
. version 10
. mata:
: function example()
: {
:         . . .
: }
: end
```

In the above, we compile example() by using the version 10 syntax of the Mata language, and any functions example() calls will be the 10 version of those functions. Setting version 10 before entering Mata ensured all of that.

In the following example, we compile using version 10 syntax and use version 10.2 functions:

```
. version 10
. mata:
: function example()
: {
:         version 10.2
:         . . .
: }
: end
```

In the following example, we compile using version 10.2 syntax and use version 10 functions:

```
. version 10.2
. mata:
: function example()
: {
:         version 10
:         . . .
: }
: end
```

It is, however, very rare that you will want to compile and execute at different version levels.

Also See

[M-5] **callersversion( )** — Obtain version number of caller

[M-2] **intro** — Language definition

Title

[M-2] void — Void matrices

Syntax

J(0, 0, .)	0×0 real matrix
J(r, 0, .)	$r \times 0$ real matrix
J(0, c, .)	$0 \times c$ real matrix
J(0, 0, 1i)	0×0 complex matrix
J(r, 0, 1i)	$r \times 0$ complex matrix
J(0, c, 1i)	$0 \times c$ complex matrix
J(0, 0, "")	0×0 string matrix
J(r, 0, "")	$r \times 0$ string matrix
J(0, c, "")	$0 \times c$ string matrix
J(0, 0, NULL)	0×0 pointer matrix
J(r, 0, NULL)	$r \times 0$ pointer matrix
J(0, c, NULL)	$0 \times c$ pointer matrix

Description

Mata allows 0×0, $r \times 0$, and $0 \times c$ matrices. These matrices are called *void matrices*.

Remarks

Remarks are presented under the following headings:

> *Void matrices, vectors, row vectors, and column vectors*
> *How to read conformability charts*

Void matrices, vectors, row vectors, and column vectors

Void matrices contain nothing, but they have dimension information (they are 0×0, $r \times 0$, or $0 \times c$) and have an *eltype* (which is real, complex, string, or pointer):

1. A matrix is said to be void if it is 0×0, $r \times 0$, or $0 \times c$.

2. A vector is said to be void if it is 0×1 or 1×0.

3. A column vector is said to be void if it is 0×1.

4. A row vector is said to be void if it is 1×0.

5. A scalar cannot be void because it is, by definition, 1×1.

The function J(r, c, *val*) creates $r \times c$ matrices containing *val*; see [M-5] **J()**. J() can be used to manufacture void matrices by specifying r and/or c as 0. The value of the third argument does not matter, but its *eltype* does:

1. J(0,0,.) creates a real 0×0 matrix, as will J(0,0,1) and as will J() with any real third argument.

2. J(0,0,1i) creates a 0×0 complex matrix, as will J() with any complex third argument.

3. J(0,0,"") creates 0×0 string matrices, as will J() with any string third argument.

4. J(0,0,NULL) creates 0×0 pointer matrices, as will J() with any pointer third argument.

In fact, one rarely needs to manufacture such matrices because they arise naturally in extreme cases. Similarly, one rarely needs to include special code to handle void matrices because such matrices handle themselves. Loops vanish when the number of rows or columns are zero.

How to read conformability charts

In general, not only is no emphasis placed on how functions and operators deal with void matrices, no mention is even made of the fact. Instead, the information is buried in the *Conformability* section located near the end of the function's or operator's manual entry.

For instance, the conformability chart for some function might read

somefunction(A, B, v):
         A: $r \times c$
         B: $c \times k$
         v: $1 \times k$ or $k \times 1$
    result: $r \times k$

Among other things, the chart above is stating how somefunction() handles void matrices. A must be $r \times c$. That chart does not say
         A: $r \times c, r > 0, c > 0$

and that is what it would have said if somefunction() did not allow A to be void. Hence, A may be 0×0, $0 \times c$, or $r \times 0$.

Similarly, B may be void as long as rows(B)==cols(A). v may be void if cols(B)==0. The returned result will be void if rows(A)==0 or cols(B)==0.

Interestingly, somefunction() can produce a nonvoid result from void input. For instance, if A were 5×0 and B, 0×3, a 5×3 result would be produced. It is interesting to speculate what would be in that 5×3 result. Probably, if we knew what somefunction() did, it would be obvious to us, but if it were not, the following section, *Diagnostics*, would state what the surprising result would be.

As a real example, see [M-5] **trace( )**. trace() will take the trace of a 0×0 matrix. The result is 0. Or see multiplication (*) in [M-2] **op_arith**. One can multiply a $k \times 0$ matrix by a $0 \times m$ matrix to produce a $k \times m$ result. The matrix will contain zeros.

Also See

[M-2] **intro** — Language definition

Title

Syntax

```
while (exp) stmt


while (exp) {
        stmts
}
```

where *exp* must evaluate to a real scalar.

Description

while executes *stmt* or *stmts* zero or more times. The loop continues as long as *exp* is not equal to zero.

Remarks

To understand while, enter the following program

```
function example(n)
{
     i = 1
     while (i<=n) {
            printf("i=%g\n", i)
            i++
     }
     printf("done\n")
}
```

and run example(3), example(2), example(1), example(0), and example(-1).

One common use of while is to loop until convergence:

```
while (mreldif(a, lasta)>1e-10) {
     lasta = a
     a = ...
}
```

Also See

[M-2] **semicolons** — Use of semicolons

[M-2] **do** — do ... while (exp)

[M-2] **for** — for (exp1; exp2; exp3) stmt

[M-2] **break** — Break out of for, while, or do loop

[M-2] **continue** — Continue with next iteration of for, while, or do loop

[M-2] **intro** — Language definition

[M-3] Commands for controlling Mata

Title

[M-3] intro — Commands for controlling Mata

Contents

Command for invoking Mata from Stata:

[M-3] Entry	Description
. mata	invoke Mata

Once you are running Mata, you can use the following commands from the colon prompt:

[M-3] Entry	Description
: mata help	execute `help` command
: mata clear	clear Mata
: mata describe	describe contents of Mata's memory
: mata memory	display memory-usage report
: mata rename	rename matrix or function
: mata drop	remove from memory matrix or function
: mata mosave	create object file
: mata mlib	create function library
: mata matsave	save matrices
: mata matuse	restore matrices
: mata matdescribe	describe contents of matrix file
: mata which	identify function
: mata query	display values of settable parameters
: mata set	set parameters
: mata stata	execute Stata command
: end	exit Mata and return to Stata

Description

When you type something at the Mata prompt, it is assumed to be a Mata statement—something that can be compiled and executed—such as

```
: 2+3
  5
```

The mata command, however, is different. When what you type is prefixed by the word mata, think of yourself as standing outside of Mata and giving an instruction that affects the Mata environment and the way Mata works. For instance, typing

```
: mata clear
```

says that Mata is to be cleared. Typing

```
: mata set matastrict on
```

says that Mata is to require that programs explicitly declare their arguments and their working variables; see [M-2] **declarations**.

Remarks

The mata command cannot be used inside functions. It would make no sense to code

```
function foo(...)
{
        ...
        mata query
        ...
}
```

because mata query is something that can be typed only at the Mata colon prompt:

```
: mata query
  (output omitted )
```

See [M-1] **how**.

Also See

[M-0] **intro** — Introduction to the Mata manual

Title

> **[M-3] end** — Exit Mata and return to Stata

Syntax

> : end

Description

end exits Mata and returns to Stata.

Remarks

When you exit from Mata back into Stata, Mata does not clear itself; so if you later return to Mata, you will be right back where you were. See [M-3] **mata**.

Also See

[M-3] **intro** — Commands for controlling Mata

Title

[M-3] **mata** — Mata invocation command

Syntax

The `mata` command documented here is for use from Stata. It is how you enter Mata. You type `mata` at a Stata dot prompt, not a Mata colon prompt.

Syntax 1	comment
`mata`	no colon following `mata`
istmt	
istmt	if an error occurs, you stay in
..	mata mode
istmt	
`end`	you exit when you type end

Syntax 1 is the best way to use Mata interactively.

Syntax 2	comment
`mata:`	colon following `mata`
istmt	
istmt	if an error occurs, you are
..	dumped from mata
istmt	
`end`	otherwise, you exit when you type end

Syntax 2 is mostly used by programmers in ado-files.
Programmers want errors to stop everything.

Syntax 3	comment
`mata` *istmt*	rarely used

Syntax 3 is the single-line variant of Syntax 1, but it is not useful.

Syntax 4	comment
`mata:` *istmt*	for use by programmers

Syntax 4 is the single-line variant of Syntax 2, and it exists for the same reason as Syntax 2: for use by programmers in ado-files.

Description

The `mata` command invokes Mata. An *istmt* is something Mata understands; *istmt* stands for interactive statement of Mata.

Remarks

Remarks are presented under the following headings:

> *Introduction*
> *The fine distinction between syntaxes 3 and 4*
> *The fine distinction between syntaxes 1 and 2*

Introduction

For interactive use, use syntax 1. Type `mata` (no colon), press *Enter*, and then use Mata freely. Type `end` to return to Stata. (When you exit from Mata back into Stata, Mata does not clear itself; so if you later type `mata`-followed-by-enter again, you will be right back where you were.)

For programming use, use Syntax 2 or Syntax 4. Inside a program or an ado-file, you can just call a Mata function

```
program myprog
    ...
    mata: utility("`varlist'")
    ...
end
```

and you can even include that Mata function in your ado-file

────────────────────────────── top of `myprog.ado` ──────────

```
program myprog
    ...
    mata: utility("`varlist'")
    ...
end

mata:
function utility(string scalar varlist)
{
    ...
}
end
```

────────────────────────────── end of `myprog.ado` ──────────

or you could separately compile `utility()` and put it in a `.mo` file or in a Mata library.

The fine distinction between syntaxes 3 and 4

Syntaxes 3 and 4 are both single-line syntaxes. You type `mata`, perhaps a colon, and follow that with the Mata *istmt*.

The differences between the two syntaxes is whether they allow continuation lines. With a colon, no continuation line is allowed. Without a colon, you may have continuation lines.

For instance, let's consider

```
function renorm(scalar a, scalar b)
{
        ...
}
```

No matter how long the function, it is one *istmt*. Using `mata:`, if you were to try to enter that *istmt*, here is what would happen:

```
. mata: function renorm(scalar a, scalar b)
<istmt> incomplete
r(197);
```

When you got to the end of the first line and pressed *Enter*, you got an error message. Using the `mata:` command, the *istmt* must all fit on one line.

Now try the same thing using `mata` without the colon:

```
. mata function renorm(scalar a, scalar b)
> {
>        ...
> }
.
```

That worked! Single-line `mata` without the colon allows continuation lines and, on this score at least, seems better than single-line `mata` with the colon. In programming contexts, however, this feature can bite. Consider the following program fragment:

```
program example
        ...
        mata utility("'varlist'"
        replace 'x' = ...
        ...
end
```

We used `mata` without the colon, and we made an error: we forgot the close parenthesis. `mata` without the colon will be looking for that close parenthesis and so will eat the next line—a line not intended for Mata. Here we will get an error message because "`replace 'x' = ...`" will make no sense to Mata, but that error will be different from the one we should have gotten. In the unlikely worse case, that next line will make sense to Mata.

Ergo, programmers want to include the colon. It will make your programs easier to debug.

There is, however, a programmer's use for single-line `mata` without the colon. In our sample ado-file above when we included the routine `utility()`, we bound it in `mata:` and `end`. It would be satisfactory if instead we coded

——————————————————————————— top of `myprog.ado` ———————

```
program myprog
        ...
        mata: utility("'varlist'")
        ...
end

mata function utility(string scalar varlist)
{
        ...
}
```

——————————————————————————— end of `myprog.ado` ———————

Using `mata` without the colon, we can omit the `end`. We admit we sometimes do that.

The fine distinction between syntaxes 1 and 2

Nothing said above about continuation lines applies to syntaxes 1 and 2. The multiline mata, with or without colon, always allows continuation lines because where the Mata session ends is clear enough: end.

The difference between the two multiline syntaxes is whether Mata tolerates errors or instead dumps you back into Stata. Interactive users appreciate tolerance. Programmers want strictness. Programmers, consider the following (using mata without the colon):

```
program example2
    ...
    mata
        result = myfunc("`varlist'")
        st_local("n" result)              /* <- mistake here */
        result = J(0,0,"")
    end
    ...
end
```

In the above example, we omitted the comma between "n" and result. We also used multiline mata without the colon. Therefore, the incorrect line will be tolerated by Mata, which will merrily continue executing our program until the end statement, at which point Mata will return control to Stata and not tell Stata that anything went wrong! This could have serious consequences, all of which could be avoided by substituting multiline mata with the colon.

Also See

[M-3] **intro** — Commands for controlling Mata

Title

> **[M-3] mata clear** — Clear Mata's memory

Syntax

> : mata clear

This command is for use in Mata mode following Mata's colon prompt. To use this command from Stata's dot prompt, type

> . mata: mata clear

Description

mata clear clears Mata's memory, in effect resetting Mata. All functions, matrices, etc., are freed.

Remarks

Stata can call Mata which can call Stata, which can call Mata, etc. In such cases, mata clear releases only resources that are not in use by prior invocations of Mata.

Stata's clear all command (see [D] **clear**) performs a mata clear, among other things.

See [M-3] **mata drop** for clearing individual matrices or functions.

Also See

[M-3] **mata drop** — Drop matrix or function

[M-3] **intro** — Commands for controlling Mata

Title

[M-3] mata describe — Describe contents of Mata's memory

Syntax

: mata <u>d</u>escribe [*namelist*] [, all]

: mata <u>d</u>escribe using *libname*

where *namelist* is as defined in [M-3] **namelists**. If *namelist* is not specified, "* *()" is assumed.

This command is for use in Mata mode following Mata's colon prompt. To use this command from Stata's dot prompt, type

. mata: mata describe ...

Description

mata describe lists the names of the matrices and functions in memory, including the amount of memory consumed by each.

mata describe using *libname* describes the contents of the specified .mlib library; see [M-3] **mata mlib**.

Option

all specifies that automatically loaded library functions that happen to be in memory are to be included in the output.

Remarks

mata describe is often issued without arguments, and then everything in memory is described:

```
: mata describe
```

# bytes	type	name and extent
50	real matrix	foo()
1,600	real matrix	X[10,20]
8	real scalar	x

mata describe using *libname* lists the functions stored in a .mlib library:

```
: mata describe using lmatabase
```

# bytes	type	name and extent
984	auto numeric vector	Corr()
340	auto real matrix	Hilbert()
(output omitted)		
528	auto transmorphic colvector	vech()

201

Diagnostics

The reported memory usage does not include overhead, which usually amounts to 64 bytes, but can be less (as small as zero for recently used scalars).

The reported memory usage in the case of pointer matrices reflects the memory used to store the matrix itself and does not include memory consumed by siblings.

Also See

[M-5] **sizeof( )** — Number of bytes consumed by object

[M-3] **intro** — Commands for controlling Mata

Title

Syntax

: mata drop *namelist*

where *namelist* is as defined in [M-3] **namelists**.

This command is for use in Mata mode following Mata's colon prompt. To use this command from Stata's dot prompt, type

. mata: mata drop ...

Description

mata drop clears from memory the specified matrices and functions.

Remarks

Use mata describe (see [M-3] **mata describe**) to determine what is in memory. Use mata clear (see [M-3] **mata clear**) to drop all matrices and functions, or use Stata's clear mata command (see [D] **clear**).

To drop a matrix named A, type

: mata drop A

To drop a function named foo(), type

: mata drop foo()

To drop a matrix named A and a function named foo(), type

: mata drop A foo()

Also See

[M-3] **mata clear** — Clear Mata's memory

[M-3] **intro** — Commands for controlling Mata

Title

[M-3] **mata help** — Obtain online help

Syntax

: mata help ...

: help ...

help need not be preceded by mata.

Description

mata help is Stata's help command. Thus you do not have to exit Mata to use help.

Remarks

See [M-1] **help**.

You need not type the Mata prefix:

: mata help mata
(output appears in Stata's Viewer)
: help mata
(same result)

help can be used to obtain help for Mata or Stata:

: help mata missing()
: help missing

Also See

[M-1] **help** — Obtaining online help

[R] **help** — Display online help

[M-3] **intro** — Commands for controlling Mata

Title

> **[M-3] mata matsave** — Save and restore matrices

Syntax

> : mata matsave *filename namelist* [, replace]
>
> : mata matuse *filename* [, replace]
>
> : mata <u>mat</u>describe *filename*

where *namelist* is a list of matrix names as defined in [M-3] **namelists**.

If *filename* is specified without a suffix, .mmat is assumed.

These commands are for use in Mata mode following Mata's colon prompt. To use these commands from Stata's dot prompt, type

> . mata: mata matsave ...

Description

mata matsave saves the specified global matrices in *filename*.

mata matuse loads the matrices stored in *filename*.

mata matdescribe describes the contents of *filename*.

Option for mata matsave

replace specifies that *filename* may be replaced if it already exists.

Option for mata matuse

replace specifies that any matrices in memory with the same name as those stored in *filename* can be replaced.

Remarks

These commands are for interactive use; they are not for use inside programs. See [M-5] **fopen( )** for Mata's programming functions for reading and writing files. In the programming environment, if you have a matrix X and want to write it to file mymatrix.myfile, you code

```
fh = fopen("mymatrix.myfile", "w")
fputmatrix(fh, X)
fclose(fh)
```

Later, you can read it back by coding

```
fh = fopen("mymatrix.myfile", "r")
X = fgetmatrix(fh)
fclose(fh)
```

mata matsave, mata matuse, and mata matdescribe are for use outside programs, when you are working interactively. You can save your global matrices

```
: mata matsave mywork *
(saving A, X, Z, beta)
file mywork.mmat saved
```

and then later get them back:

```
: mata matuse mywork
(loading A, X, Z, beta)
```

mata matdescribe will tell you the contents of a file:

```
: mata matdescribe mywork
file mywork.mmat saved on 4 Apr 2007 08:46:39 contains
X, X, Z, beta
```

Diagnostics

mata matsave saves the contents of view matrices. Thus when they are restored by mata matuse, the contents will be correct regardless of the data Stata has loaded in memory.

Also See

[M-3] **intro** — Commands for controlling Mata

Title

[M-3] **mata memory** — Report on Mata's memory usage

Syntax

```
: mata memory
```

This command is for use in Mata mode following Mata's colon prompt. To use this command from Stata's dot prompt, type

```
. mata: mata memory
```

Description

`mata memory` provides a summary of Mata's current memory utilization.

Remarks

```
: mata memory
```

	#	bytes
autoloaded functions	15	5,514
ado functions	0	0
defined functions	0	0
matrices & scalars	14	8,256
overhead		1,972
total	29	15,742

Functions are divided into three types:

1. *Autoloaded functions*, which refer to library functions currently loaded (and which will be automatically unloaded) from `.mlib` library files and `.mo` object files.

2. *Ado-functions*, which refer to functions currently loaded whose source appears in ado-files themselves. These functions will be cleared when the ado-file is automatically cleared.

3. *Defined functions*, which refer to functions that have been defined either interactively or from do-files and which will be cleared only when a `mata clear`, `mata drop`, Stata `clear mata`, or Stata `clear all` command is executed; see [M-3] **mata clear**, [M-3] **mata drop**, or [D] **clear**.

Also See

[M-1] **limits** — Limits and memory utilization

[M-3] **mata clear** — Clear Mata's memory

[D] **memory** — Memory size considerations

[M-3] **intro** — Commands for controlling Mata

Title

> **[M-3] mata mlib** — Create function library

Syntax

> : mata mlib create *libname*   [, dir(*path*) replace]
>
> : mata mlib add *libname fcnlist*() [, dir(*path*)]
>
> : mata mlib index
>
> : mata mlib <u>q</u>uery

where *fcnlist*() is a *namelist* containing only function names, such as

fcnlist() examples
myfunc()
myfunc() myotherfunc() foo()
f*() g*()
*()

see [M-3] **namelists**

and where *libname* is the name of a library. You must start *libname* with the letter l and do not add the .mlib suffix as it will be added for you. Examples of *libnames* include

libname	corresponding filename
lmath	lmath.mlib
lmoremath	lmoremath.mlib
lnjc	lnjc.mlib

Also *libnames* that begin with the letters lmata, such as lmatabase, are reserved for the names of official libraries.

This command is for use in Mata mode following Mata's colon prompt. To use this command from Stata's dot prompt, type

> . mata: mata mlib ...

Description

mata mlib creates, adds to, and causes Mata to index .mlib files, which are libraries containing the object-code functions.

mata mlib create creates a new, empty library.

mata mlib add adds new members to a library.

mata mlib index causes Mata to build a new list of libraries to be searched.

mata mlib query lists the libraries to be searched.

A library may contain up to 500 functions.

Options

dir(*path*) specifies the directory (folder) into which the file should be written. dir(.) is the default, meaning that if dir() is not specified, the file is written into the current (working) directory. *path* may be a directory name or may be the sysdir shorthand STATA, UPDATES, BASE, SITE, PLUS, PERSONAL, or OLDPLACE; see [P] **sysdir**. dir(PERSONAL) is recommended.

replace specifies that the file may be replaced if it already exists.

Remarks

Remarks are presented under the following headings:

> *Background*
> *Outline of the procedure for dealing with libraries*
> *Creating a .mlib library*
> *Adding members to a .mlib library*
> *Listing the contents of a library*
> *Making it so Mata knows to search your libraries*
> *Advice on organizing your source code*

Also see [M-1] **how** for an explanation of object code.

Background

.mlib files contain the object code for one or more functions. Functions which happen to be stored in libraries are called library functions, and Mata's library functions are also stored in .mlib libraries. You can create your own libraries, too.

Mata provides two ways to store object code:

1. In a .mo file, which contains the code for one function

2. In a .mlib library file, which may contain the code for up to 500 functions

.mo files are easier to use and work just as well as .mlib libraries; see [M-3] **mata mosave**. .mlib libraries, however, are easier to distribute to others when you have many functions, because they are combined into one file.

Outline of the procedure for dealing with libraries

Working with libraries is easy:

1. First, choose a name for your library. We will choose the name lpersonal.

2. Next, create an empty library by using the `mata mlib create` command.

3. After that, you can add new members to the library at any time, using `mata mlib add`.

`.mlib` libraries contain object code, not the original source code, so you need to keep track of the source code yourself. Also, if you want to update the object code in a function stored in a library, you must recreate the entire library; there is no way to replace or delete a member once it is added.

We begin by showing you the mechanical steps, and then we will tell you how we manage libraries and source code.

Creating a .mlib library

If you have not read [M-3] **mata mosave**, please do so.

To create a new, empty library named `lpersonal.mlib` in the current directory, type

```
: mata mlib create lpersonal
file lpersonal.mlib created
```

If `lpersonal.mlib` already exists and you want to replace it, either erase the existing file first or type

```
: mata mlib create lpersonal, replace
file lpersonal.mlib created
```

To create a new, empty library named `lpersonal.mlib` in your PERSONAL (see [P] **sysdir**) directory, type

```
: mata mlib create lpersonal, dir(PERSONAL)
file c:\ado\personal\lpersonal.mlib created
```

or

```
: mata mlib create lpersonal, dir(PERSONAL) replace
file c:\ado\personal\lpersonal.mlib created
```

Adding members to a .mlib library

Once a library exists, whether you have just created it and it is empty, or it already existed and contains some functions, you can add new functions to it by typing

```
: mata mlib add libname fcnname()
```

So, if we wanted to add function `example()` to library `lpersonal.mlib`, we would type

```
: mata mlib add lpersonal example()
(1 function added)
```

In doing this, we do not have to say where `lpersonal.mlib` is stored; Mata searches for it along the ado-path.

Before you can add `example()` to the library, however, you must compile it:

```
: function example(...)
> {
>        ...
> }

: mata mlib add lpersonal example()
(1 function added)
```

You can add many functions to a library in one command:

```
: mata mlib add lpersonal example2() example3()
(2 functions added)
```

You can add all the functions currently in memory by typing

```
: mata mlib add lanother *()
(3 functions added)
```

In the above example, we added to `lanother.mlib` because we had already added `example()`, `example2()`, and `example3()` to `lpersonal.mlib` and trying to add them again would result in an error. (Before adding `*()`, we could verify that we are adding what we want to add by typing `mata describe *()`; see [M-3] **mata describe**.)

Listing the contents of a library

Once a library exists, you can list its contents (the names of the functions it contains) by typing

```
: mata describe using libname
```

Here we would type

```
: mata describe using lpersonal
(library contains 3 members)
```

# bytes	type	name and extent
32	auto transmorphic matrix	example()
32	auto transmorphic matrix	example2()
32	auto transmorphic matrix	example3()

`mata describe` usually lists the contents of memory, but `mata describe using` lists the contents of a library.

Making it so Mata knows to search your libraries

Mata automatically finds the `.mlib` libraries on your computer. It does this when Mata is invoked for the first time during a session. Thus everything is automatic except that Mata will know nothing about any new libraries created during the Stata session, so after creating a new library, you must tell Mata about it. You do this by asking Mata to rebuild its library index:

```
: mata mlib index
```

You do not specify the name of your new library. That name does not matter because Mata rebuilds its entire library index.

You can issue the `mata mlib index` command right after creating the new library

```
: mata mlib create lpersonal, dir(PERSONAL)
: mata mlib index
```

or after you have created and added to the library:

```
: mata mlib create lpersonal, dir(PERSONAL)
: mata mlib add lpersonal *()
: mata mlib index
```

It does not matter. Mata does not need to rebuild its index after a known library is updated; Mata needs to be told to rebuild only when a new library is added during the session.

Advice on organizing your source code

Say you wish to create and maintain `lpersonal.mlib`. Our preferred way is to use a do-file:

——————————————————————————————————————— top of `lpersonal.do` ———————

```
mata:
mata clear
```
function definitions appear here
```
mata mlib create lpersonal, dir(PERSONAL) replace
mata mlib add lpersonal *()
mata mlib index
end
```

——————————————————————————————————————— end of `lpersonal.do` ———————

This way, all we have to do to create or recreate the library is enter Stata, change to the directory containing our source code, and type

```
. do lpersonal
```

For large libraries, we like to put the source code for different parts in different files:

——————————————————————————————————————— top of `lpersonal.do` ———————

```
mata: mata clear
do function1.mata
do function2.mata
...
mata:
mata mlib create lpersonal, dir(PERSONAL) replace
mata mlib add lpersonal *()
mata mlib index
end
```

——————————————————————————————————————— end of `lpersonal.do` ———————

The function files contain the source code, which might include one function, or it might include more than one function if the primary function had subroutines:

——————————————————————————————————————— top of `function1.mata` ———————

```
mata:
```
function definitions appear here
```
end
```

——————————————————————————————————————— end of `function1.mata` ———————

We name our component files ending in `.mata`, but they are still just do-files.

Also See

[M-3] **mata mosave** — Save function's compiled code in object file

[M-3] **intro** — Commands for controlling Mata

Title

[M-3] mata mosave — Save function's compiled code in object file

Syntax

> : mata mosave *fcname*() $\left[\,,\ \text{dir}(path)\ \text{replace}\,\right]$

This command is for use in Mata mode following Mata's colon prompt. To use this command from Stata's dot prompt, type

> . mata: mata mosave ...

Description

mata mosave saves the object code for the specified function in the file *fcnname*.mo.

Options

dir(*path*) specifies the directory (folder) into which the file should be written. dir(.) is the default, meaning that if dir() is not specified, the file is written into the current (working) directory. *path* may be a directory name or may be the sysdir shorthand STATA, UPDATES, BASE, SITE, PLUS, PERSONAL, or OLDPLACE; see [P] **sysdir**. dir(PERSONAL) is recommended.

replace specifies that the file may be replaced if it already exists.

Remarks

See [M-1] **how** for an explanation of object code.

Remarks are presented under the following headings:

> *Example of use*
> *Where to store .mo files*
> *Use of .mo files versus .mlib files*

Example of use

.mo files contain the object code for one function. If you store a function's object code in a .mo file, then in future Mata sessions, you can use the function without recompiling the source. The function will appear to become a part of Mata just as all the other functions documented in this manual are. The function can be used because the object code will be automatically found and loaded when needed.

For example,

```
: function add(a,b) return(a+b)
: add(1,2)
  3
```

```
: mata mosave add()
(file add.mo created)
: mata clear
: add(1,2)
  3
```

In the example above, function add() was saved in file add.mo stored in the current directory. After clearing Mata, we could still use the function because Mata found the stored object code.

Where to store .mo files

Mata could find add() because file add.mo was in the current directory, and our ado-path included .:

```
. adopath
  [1]  (UPDATES)    "C:\Program Files\Stata10\ado\updates\"
  [2]  (BASE)       "C:\Program Files\Stata10\ado\base\"
  [3]  (SITE)       "C:\Program Files\Stata10\ado\site\"
  [4]               "."
  [5]  (PERSONAL)   "C:\ado\personal\"
  [6]  (PLUS)       "C:\ado\plus\"
  [7]  (OLDPLACE)   "C:\ado\"
```

If later we were to change our current directory

```
. cd ..\otherdir
```

Mata would no longer be able to find the file add.mo. Thus the best place to store your personal .mo files is in your PERSONAL directory. Thus rather than typing

```
: mata mosave example()
```

we would have been better off typing

```
: mata mosave example(), dir(PERSONAL)
```

Use of .mo files versus .mlib files

Use of .mo files is heartily recommended. The alternative for saving compiled object code are .mlib libraries; see [M-3] **mata mlib** and [M-1] **ado**.

Libraries are useful when you have many functions and want to tie them together into one file, especially if you want to share those functions with others, because then you have only one file to distribute. The disadvantage of libraries is that you must rebuild them whenever you wish to remove or change the code of one. If you have only a few object files, or if you have many but sharing is not an issue, .mo libraries are easier to manage.

Also See

[M-3] **mata mlib** — Create function library

[M-3] **intro** — Commands for controlling Mata

Title

[M-3] **mata rename** — Rename matrix or function

Syntax

> : mata rename *oldmatrixname newmatrixname*

> : mata rename *oldfcnname*() *newfcnname*()

This command is for use in Mata mode following Mata's colon prompt. To use this command from Stata's dot prompt, type

> . mata: mata rename ...

Description

mata rename changes the names of functions and global matrices.

Also See

[M-3] **intro** — Commands for controlling Mata

Title

[M-3] **mata set** — Set and display Mata system parameters

Syntax

```
: mata query

: mata set matacache     # [ , permanently ]

: mata set matalnum      { off | on }

: mata set mataoptimize  { on | off }

: mata set matafavor     { space | speed } [ , permanently ]

: mata set matastrict    { off | on } [ , permanently ]

: mata set matalibs      "libname ; libname ; ... "

: mata set matamofirst   { off | on } [ , permanently ]
```

These commands are for use in Mata mode following Mata's colon prompt. To use these commands from Stata's dot prompt, type

```
. mata: mata query

. mata: mata set ...
```

Description

`mata query` shows the values of Mata's system parameters.

`mata set` sets the value of the system parameters:

`mata set matacache` specifies the maximum amount of memory, in kilobytes, that may be consumed before Mata starts looking to drop autoloaded functions that are not currently being used. The default value is 400, meaning 400 kilobytes. This parameter affects the efficiency with which Stata runs. Larger values cannot hurt, but once `matacache` is large enough, larger values will not improve performance.

`mata set matalnum` turns program line-number tracing on or `off`. The default setting is `off`. This setting modifies how programs are compiled. Programs compiled when `matalnum` is turned on include code so that, if an error occurs during execution of the program, the line number is also reported. Turning `matalnum` on prevents Mata from being able to optimize programs, so they will run more slowly. Except when debugging, the recommended setting for this is `off`.

`mata set mataoptimize` turns compile-time code optimization on or `off`. The default setting is `on`. Programs compiled when `mataoptimize` is switched off will run more slowly and, sometimes, much more slowly. The only reason to set `mataoptimize` off is if a bug in the optimizer is suspected.

mata set matafavor specifies whether, when executing code, Mata should favor conserving
memory (space) or running quickly (speed). The default setting is space. Switching to
speed will make Mata, in a few instances, run a little quicker but consume more memory.
Also see [M-5] **favorspeed( )**.

mata set matastrict sets whether declarations can be omitted inside the body of a program. The
default is off. If matastrict is switched on, compiling programs that omit the declarations
will result in a compile-time error; see [M-2] **declarations**. matastrict acts unexpectedly but
pleasingly when set/reset inside ado-files; see [M-1] **ado**.

mata set matalibs sets the names and order of the .mlib libraries to be searched; see
[M-1] **how**. matalibs usually is set to "lmatabase;lmataopt;lmataado". However it is
set, it is probably set correctly, because Mata automatically searches for libraries the first time
it is invoked in a Stata session. If, during a session, you erase or copy new libraries along
the ado-path, the best way to reset matalibs is with the mata mlib index command; see
[M-3] **mata mlib**. The only reason to set matalibs by hand is to modify the order in which
libraries are searched.

mata set matamofirst states whether .mo files or .mlib libraries are searched first. The default
is off, meaning libraries are searched first.

Option

permanently specifies that, in addition to making the change right now, the setting be remembered
and become the default setting when you invoke Stata in the future.

Remarks

Remarks are presented under the following headings:

> *Relationship between Mata's mata set and Stata's set commands*
> *c( ) values*

Relationship between Mata's mata set and Stata's set commands

The command

        : mata set ...

issued from Mata's colon prompt and the command

        . set ...

issued from Stata's dot prompt are the same command, so you may set Mata's (or even Stata's)
system parameters either way.

The command

        : mata query

issued from Mata's colon prompt and the command

        . query mata

issued from Stata's dot prompt are also the same command.

c( ) values

The following concerns Stata more than Mata.

Stata's c-class, c(), contains the values of system parameters and settings along with certain other constants. c() values may be referred to in Stata, either via macro substitution ('c(current_date)', for example) or in expressions (in which case the macro quoting characters may be omitted). Stata's c() is also available in Mata via Mata's c() function; see [M-5] **c( )**.

Most everything set by set is available via c(), including Mata's set parameters:

     c(matacache) returns a numeric scalar equal to the cache size.

     c(matalnum) returns a string equal to "on" or "off".

     c(mataoptimize) returns a string equal to "on" or "off".

     c(matafavor) returns a string equal to "space" or "speed".

     c(matastrict) returns a string equal to "on" or "off".

     c(matalibs) returns a string of library names separated by semicolons.

     c(matamofirst) returns a string equal to "on" or "off".

The above is in Stataspeak. Rather than referring to c(matacache), we would refer to c("matacache") if we were using Mata's function. The real use of these values, however, is in Stata.

Also See

[R] **query** — Display system parameters

[R] **set** — Overview of system parameters

[P] **creturn** — Return c-class values

[M-3] **intro** — Commands for controlling Mata

Title

[M-3] **mata stata** — Execute Stata command

Syntax

: mata stata *stata_command*

This command is for use in Mata mode following Mata's colon prompt.

Description

mata stata *stata_command* passes *stata_command* to Stata for execution.

Remarks

mata stata is a convenience tool to keep you from having to exit Mata:

```
: st_view(V=., 1\5, ("mpg", "weight"))
no variables defined
            st_view():  3500  invalid Stata variable name
              <istmt>:     -   function returned error
r(3500);
: mata stata sysuse auto
  (1978 Automobile Data)
: st_view(V=., 1\5, ("mpg", "weight"))
```

mata stata is for interactive use. If you wish to execute a Stata command from a function, see [M-5] **stata( )**.

Also See

[M-5] **stata( )** — Execute Stata command

[M-3] **intro** — Commands for controlling Mata

Title

Syntax

> : mata which *fcnname* ( )

This command is for use in Mata mode following Mata's colon prompt. To use this command from Stata's dot prompt, type

> . mata: mata which ...

Description

mata which *fcnname* looks for *fcnname*() and reports whether it is built in, stored in a .mlib library, or stored in a .mo file.

Remarks

mata which *fcnname*() looks for *fcnname*() and reports where it is found:

```
: mata which I()
  I():  built-in
: mata which assert()
  assert():  lmatabase
: mata which myfcn()
  userfunction():  .\myfcn.mo
: mata which nosuchfunction()
function nosuchfunction() not found
r(111);
```

Function I() is built in; it was written in C and is a part of Mata itself.

Function assert() is a library function and, as a matter of fact, its executable object code is located in the official function library lmatabase.mlib.

Function myfcn() exists and has its executable object code stored in file myfcn.mo, located in the current directory.

Function nosuchfunction() does not exist.

Going back to mata which assert(), which was found in lmatabase.mlib, if you wanted to know where lmatabase.mlib was stored, you could type findfile lmatabase.mlib at the Stata prompt; see [P] **findfile**.

Also See

[M-3] **intro** — Commands for controlling Mata

Title

[M-3] namelists — Specifying matrix and function names

Syntax

Many mata commands allow or require a *namelist*, such as

: mata describe [*namelist*] [, detail all]

A *namelist* is defined as a list of matrix and/or function names, such as

alpha beta foo()

The above *namelist* refers to the matrices alpha and beta along with the function named foo().

Function names always end in (), hence

alpha	refers to the matrix named alpha
alpha()	refers to the function of the same name

Names may also be specified using the * and ? wildcard characters:

*	means zero or more characters go here
?	means exactly one character goes here

hence,

*	means all matrices
*()	means all functions
* *()	means all matrices and all functions
s*	means all matrices that start with *s*
s*()	means all functions that start with *s*
*e	means all matrices that end with *e*
*e()	means all functions that end with *e*
s*e	means all matrices that start with *s* and end with *e*
s*e()	means all functions that start with *s* and end with *e*
s?e	means all matrices that start with *s* and end with *e* and have one character in between
s?e()	means all functions that start with *s* and end with *e* and have one character in between

Description

Namelists appear in syntax diagrams.

Remarks

Some *namelists* allow only matrices, and some allow only functions. Even when only functions are allowed, you must include the () suffix.

Also See

[M-3] **intro** — Commands for controlling Mata